D0594048

Adrenaline

Adrenaline

BRIAN B. HOFFMAN

Harvard University Press

CAMBRIDGE, MASSACHUSETTS

LONDON, ENGLAND

2013

LIBRARY
FRANKLIN PIERCE UNIVERSITY
RINDGE, NH 03461

For my wife, Maggie, and our children—
Jonathan, Rachel, and Daniel

Copyright © 2013 by the President and Fellows of Harvard College
All rights reserved
Printed in the United States of America
Page 282 constitutes an extension of the copyright page.

Many of the designations used by manufacturers and sellers to
distinguish their products are claimed as trademarks. Where
those designations appear in this book and Harvard University
Press was aware of a trademark claim, the designations
have been printed in initial capital letters.

Library of Congress Cataloging-in-Publication Data

Hoffman, Brian B.
Adrenaline / Brian B. Hoffman.
p. cm.
Includes bibliographical references and index.
ISBN 978-0-674-05088-4 (alk. paper)
I. Title.
[DNLM: 1. Epinephrine—history. 2. History, 19th Century.
3. History, 20th Century. 4. History, 21st Century. WK 11.1]
616.4′5—dc23 2012035120

Book design by Dean Bornstein

Contents

Prologue

A little learning is a dangerous thing;
drink deep, or taste not the Pierian spring.

ALEXANDER POPE

Adrenaline has enormous cultural significance as a molecule associated with medically important stress and with excitement, anger, and terror. Adrenaline *flows* in an adventurer ascending a sheer cliff or in a young violinist debuting at Carnegie Hall. A *shot* of adrenaline empowers superhuman feats in emergencies. Risk takers seeking dangerous recreations are adrenaline *junkies* who enjoy repetitive adrenaline *rushes*. On the other hand, an adrenaline *surge* during a bitter argument may precipitate a heart attack or scare us to death.

Many widely held views about adrenaline are popular extrapolations of scientific research done by Walter Cannon at Harvard Medical School. Almost a hundred years ago, Cannon incorporated knowledge of adrenaline's effects on individual organs into an integrated view of physiological changes that accompany fear and rage. Cannon postulated that fight-or-flight responses evolved in our remote ancestors' brutish environments, providing a survival advantage in dangerous confrontations. This framework has proved very fruitful in understanding the importance of adrenaline in driving the heart to beat harder and faster, in opening the lungs to airflow, in stimulating the liver to pour fuel (glucose) into the bloodstream, and in facilitating many other effects that contribute to our capacity to attack or to evade capture.

The discovery of adrenaline—several decades before Cannon's work—marked a major biomedical breakthrough. The seminal importance of adrenaline as a muse that stimulated many fundamental

I

discoveries has not been adequately recognized. Establishing the mechanisms of adrenaline's actions has led to many grand discoveries of fundamental biomedical importance. The scope of these discoveries is breathtaking and provides a strong foundation for understanding the actions of many other hormones and drugs.

The story of adrenaline reveals much about medicine's transformation from a profession with a weak intellectual foundation into a scientifically based discipline involving an increasingly fruitful flow of information back and forth between scientists and clinicians. Adrenaline moved almost immediately from the laboratory into the clinic, where it was used by physicians to treat multiple diseases. Some of these treatments proved effective, while others decidedly missed the mark. The synthesis of chemicals related to adrenaline led to drugs that are more effective and safer than adrenaline itself. Some of these drugs work by mimicking adrenaline's actions, while others block the effects of our own endogenous adrenaline. Adrenaline and related drugs have major benefits in patients with diseases such as asthma, coronary artery disease, and hypertension.

This book is a biography of adrenaline, from its earliest ancestry to its discovery at the beginning of the twentieth century and through its first hundred years. This account provides a wonderful opportunity to describe many developments that have shaped our current understanding of health and disease. This historical approach touches on events that provide insight into the successes of research in physiology, pharmacology, cell biology, and knowledge of the mechanisms of disease at the molecular level. The story illustrates the major changes that have occurred in the ethics of clinical investigation, the requirements for successful drug development, and the interconnections between fundamental science and new therapeutic agents, as well as overly optimistic claims about new drugs made by both the medical profession and the popular press. Coupled with morsels about important and occasionally brilliant scientists who have worked on adrenaline, the entire story illustrates the many twists and turns, false paths, and unexpected results that characterize challenging biomedical research.

Most of my own research has focused on questions involving the actions of adrenaline and related drugs in cells and in people. Quite by chance, I stumbled upon some fascinating old manuscripts that triggered a keen interest in the history of adrenaline. I realized that the biography of adrenaline provides a revealing window into the history of scientific ideas and medical progress, as well as adrenaline's connections with emotions. Scores of intriguing stories involving adrenaline provide rich excursions into a remarkable stream of characters and disputes, demonstrating the human side of scientific discovery, industrial research, and medical practice. Many of these topics are in endnote riffs, offered as a *menu dégustation*, and are not essential to the rest of the story.

The discovery of adrenaline and subsequent biomedical breakthroughs provide the foundation for a fascinating tale involving legendary scientists. This story ranges from a nurse with murder on her mind to tumors that secrete enough adrenaline to kill, revealing, along the way, intricate and elegant molecular mechanisms that explain how many drugs work in treating serious diseases. I hope the book stimulates even the most committed adrenaline fanatic to take a deeper interest in biomedical science and the history of medicine.

[ONE]

The Goldilocks Principle

When Kristen Gilbert killed them, she used the perfect poison.

ASSISTANT U.S. ATTORNEY ARIANE VUONO, 2001

In the fairy tale about the three bears, Goldilocks tastes Papa Bear's porridge and finds it far too hot, while Mama Bear's porridge proves much too cold. Happily for Goldilocks, Baby Bear's porridge is just right. This represents an example of what's often known as the Goldilocks Principle, reflecting the desirability of an appropriate balance that avoids extremes. As we will see, the Goldilocks Principle applies as much to adrenaline as it does to porridge. In this chapter, as we begin our multifaceted discussion of adrenaline, we concentrate on the dire effects of too much or too little adrenaline, along with analogous extreme effects associated with the sympathetic nervous system.

In some nineteenth-century versions of the tale, Goldilocks herself became a meal when the bears returned; however, as the story has been retold again and again over the years, the outcome has become much more favorable for her. Analogously, diseases involving adrenaline-related extremes that were once fatal now have much better endings as a consequence of advances in biomedical research. Nonetheless, we start our tale with a story that should alarm children of all ages.

Murder She Did

Kristen Gilbert grew up in Fall River, Massachusetts, the town where Lizzie Borden was accused (but not convicted) of murdering

4

her parents with an ax in 1892. In 1989, Gilbert began her career as a young nurse at a federal hospital near Northampton in western Massachusetts. Gilbert's colleagues noticed that patients had cardiac arrests unusually frequently during her shifts. This pattern, initially attributed to sheer bad luck, gave Gilbert a macabre reputation as an "angel of death." Over time, however, the continued association between Gilbert and sudden deaths on her ward led to concern that more than just bad luck might be involved.

Colleagues began noticing unaccounted-for empty vials of adrenaline in the trash following some of the deaths. Other patients on Gilbert's ward experienced strange and dramatic increases in heart rate and blood pressure—changes that could be brought on by adrenaline. A few coworkers started to suspect that Gilbert was injecting unsuspecting patients with adrenaline, and a criminal investigation eventually led to her arrest and prosecution for multiple murders. Her motives were never fully understood, but some suggested that she committed the crimes to bring her lover, an unwitting policeman who also worked at the hospital, to her ward as alarms sounded, while others believed Gilbert sought to generate her own adrenaline highs from the frantic resuscitation efforts undertaken at the scene of each cardiac arrest.

Adrenaline is a two-edged sword for the heart. Sudden cardiac arrest kills hundreds of thousands of people each year, but some people are saved by cardiac resuscitation, which often includes adrenaline injections to increase the chance of restarting the heart. On the other hand, a large dose of adrenaline can overstimulate a normal heart, leading to fast and erratic beating that may culminate in cardiac arrest.

Using adrenaline as a murder weapon has a certain bloodless elegance, for the deaths it causes may initially appear natural. The short delay between a potentially fatal adrenaline injection and stoppage of the heart allowed Gilbert to slip out of patients' rooms before automated cardiac monitors triggered alarms. As a poison, adrenaline presents many challenges for police, toxicologists, and prosecutors, challenges that were heightened by the hospital setting of the deaths Gilbert was charged with. The stress of illness may itself cause a copious release of adrenaline from the adrenal glands, which leads to very high concentrations in the blood. Plus, patients often receive large doses of adrenaline during cardiac resuscitation. Consequently, it is very difficult to produce compelling forensic evidence that an illicit injection of adrenaline was what stopped the heart in the first place. In Gilbert's case, the prosecution won without directly demonstrating that the dead patients had excessive quantities of adrenaline in their bodies.

During her trial, a cardiologist serving as an expert witness for the prosecution testified that he believed the deaths had been caused by illicit injections of adrenaline. This testimony, along with other evidence more typically exploited in murder trials, provided sufficient grounds for the prosecution to convince the jury that Gilbert had murdered the patients. In March 2001, Gilbert received a life sentence for the murder of four patients and attempted murder of two others. She may have caused dozens more deaths.[1]

Pheochromocytomas: Deadly Chemical Factories inside the Body

In 1884, eighteen-year-old Minna Roll died in southeastern Germany about a year after developing a set of unusual symptoms. Her illness began on a winter evening when she suddenly noticed heart palpitations followed by a severe headache and feelings of intense anxiety; the episode lasted only a few minutes. She had two similar attacks over the next six months but otherwise felt well. She then developed frequent headaches, vomiting, and difficulty seeing; she soon became too sick to continue her farm work. She was admitted to a hospital, where physicians noted a rapid heart rate that episodically increased to 160–180 beats per minute. Her arteries felt tense and were hard to compress; these changes suggest severe hypertension, but her blood pressure is unknown since blood pressure was not routinely measured in clinical settings until at least ten years later. Ophthalmological examination demonstrated bleeding in the back of her eyes and swollen optic nerves, signs that in retrospect were likely due to severe hypertension. In a matter of weeks she was dead. The autopsy demonstrated damage in multiple organs, including her heart and kidneys; these findings, too, likely represent the consequences of severe hypertension. The pathologist unexpectedly found a tumor in each adrenal gland. Dr. Felix Fränkel published a detailed case record; he did not know the explanation for Minna Roll's symptoms and could only speculate about a possible connection with the tumors found at autopsy.[2]

Over the next several decades many physicians published case reports describing similar patients who had recurrent attacks of rapid heart rate, headaches, intense perspiration, and other symptoms, coupled with adrenal tumors found at autopsy. These tumors, often less than two inches in diameter, typically arose from chromaffin cells in the adrenal medulla. The tumors soon became known as pheochromocytomas.[3] However, the possible connection between the symptoms and the tumors remained obscure. In 1922, Marcel Labbé and colleagues in Paris focused attention on the connection between attacks of symptoms and simultaneous rises in blood pressure in patients who had pheochromocytomas found at autopsy; they proposed that the tumors caused the sharp rises in blood pressure.

In 1926, Charles Mayo—one of the founders of the Mayo Clinic in Rochester, Minnesota—became the first surgeon in the United

States to remove a pheochromocytoma from a living patient.[4] He operated on Mother Mary Joachim, a thirty-year-old Roman Catholic nun from Ontario, Canada. For more than a year the nun had had troubling attacks of headaches, shortness of breath, and rapid heart rate. These spells occurred frequently, lasting from thirty minutes to several hours; her baffled general physician referred her to the Mayo Clinic for further evaluation. At the onset of an attack, Mother Mary's systolic blood pressure suddenly jumped from a normal value of 120 millimeters of mercury (mm Hg) to 320 mm Hg or more, sometimes associated with transient heart failure. The attacks ebbed spontaneously, with her blood pressure returning to normal until the next paroxysm.

Mayo wondered if periodic overactivity of the sympathetic nervous system caused the attacks, as activation of the sympathetic nervous system was known to raise blood pressure. In the 1920s, the absence of effective antihypertensive drugs made severe hypertension both dangerous and exceedingly difficult to treat; surgeons sometimes operated to cut sympathetic nerves in a desperate effort to lower blood pressure. Because Mother Mary also had intermittent abdominal pain, Mayo decided to cut the sympathetic nerves in her abdomen. After opening her abdomen, however, Mayo unexpectedly found a large tumor pressing against the top of her left kidney; he focused the operation on removing the tumor—later identified as a pheochromocytoma—and then closed the incision. Remarkably, after removal of the tumor, Mother Mary Joachim had no further episodes of high blood pressure, and lived for another eighteen years without experiencing another attack. Removing the pheochromocytoma had cured her.

In 1929, Maurice Pincoffs in Baltimore likely made the first preoperative diagnosis of a pheochromocytoma. His patient was a twenty-five-year-old woman with increasingly severe paroxysms of a racing pulse that had started ten years earlier. Aside from these attacks, her health was excellent. Pincoffs detected no abnormalities in his initial physical examination, and he noted that her blood pressure was a perfectly normal 120/80 mm Hg. He later examined her during several attacks; on those occasions her systolic blood pressure was at least 260 mm Hg. She looked very anxious and tremulous, and her blood sugar became abnormally high during attacks. Although X-rays of the abdomen were normal and he had no laboratory tests to guide him, Pincoffs astutely diagnosed a pheochromocytoma.

The patient was referred to the Baltimore surgeon Arthur Shipley, who decided to operate even though he could not be sure that she had a pheochromocytoma or in which adrenal gland the tumor might be located. He decided against opening the middle of her abdomen with a large incision, even though this approach would allow access to a tumor in either the left or right adrenal gland, as he was not confident that the patient would survive such a major procedure. Instead, he decided to make a small incision on one side; if he guessed wrong about the location of the tumor, he planned to allow her to recover for several weeks and then repeat the surgery on the other side.

Shipley made an incision on her left side and identified the normal left adrenal gland. However, sliding his arm through the incision, he felt a tumor in the right adrenal gland. At the second operation weeks later, he successfully removed the pheochromocytoma from the right adrenal gland; the patient's blood pressure dropped precipitously after the tumor was removed, but she survived the surgery. Extracts of this tumor subsequently were shown to contain considerable amounts of adrenaline (the discovery of adrenaline three decades earlier is described in Chapters 3 and 4).

By 1930, several investigators had reported that other pheochromocytomas obtained at autopsy contained large amounts of adrenaline. In 1937, investigators in New York demonstrated the presence of a blood-pressure-raising chemical similar to adrenaline in the blood of a patient with pheochromocytoma.[5] Over the next decade, new diagnostic tests facilitated the diagnosis of pheochromocytoma and helped avoid futile major surgery in patients who did not actually harbor a tumor.[6] Nonetheless, at the beginning of the 1950s, diagnosing a pheochromocytoma remained difficult, and localizing the tumor prior to surgery was often impossible, as routine X-rays were normal in most patients. Moreover, surgery could be dangerous due to wild swings in blood pressure and heart rate on the operating table.

Since that time, numerous advances have transformed care of patients with suspected pheochromocytoma. In the first place, making a diagnosis of pheochromocytoma has become much easier thanks to the development of increasingly sensitive and specific assays for adrenaline, noradrenaline, and their breakdown products. Innovative tests can now confirm or exclude the diagnosis of pheochromocytoma with considerable reliability. And because these tumors can run in families, sometimes in syndromes associated with tumors in other glands, modern genetic evaluations can be very useful in predicting the risk of pheochromocytoma in members of a family based on information involving mutations in a small number of genes. In 2007, in fact, scientists reported success in tracking down descendants in Minna Roll's extended family more than a hundred years after her death and determined that some of these relatives had inherited a mutation in a single gene that causes pheochromocytomas.[7]

In about 90 percent of cases, pheochromocytomas are found in one adrenal or the other. Modern CT or MRI scans localize tumors with considerable precision with images unavailable to physicians for much of the twentieth century. Tumors outside the adrenal glands, which can be in the neck, chest, abdomen, or pelvis, can also be tracked down using radioactive substances that are preferentially taken up by pheochromocytomas. A deeper understanding of the physiological consequences of excess adrenaline and noradrenaline has made anesthesia of these patients much safer, leading to improved outcomes. Moreover, highly specific drugs that prevent many of adrenaline's effects have improved the treatment of high blood pressure and rapid heart rate prior to surgery.

Despite these enormous advances in the biological and clinical understanding of pheochromocytoma, patients often have symptoms for months or years before a clinician makes the correct diagnosis— or the undiagnosed patient dies of a complication. Because this tumor is so rare, it is often missed. A typical general physician might only diagnose a few cases during his or her career; indeed, many physicians encounter substantially more patients with odd spells that are suggestive of pheochromocytoma but who do not have this tumor. This syndrome, incidentally, has been called pseudo-pheochromocytoma, and it represents a diagnosis of exclusion that is sometimes associated with panic or repressed emotional responses. Even more puzzling and challenging are individuals who pretend to have a pheochromocytoma. People have faked the symptoms, injecting themselves with adrenaline or surreptitiously added it to their urine to falsify laboratory tests. Some of these people have even willingly undergone surgery intended to remove a tumor they don't have. This behavior constitutes a form of Munchausen's syndrome, a disorder in which affected people feign illness for no apparent reason.[8]

Two well-known people who died with an undiagnosed pheochromocytoma are Pauline Hemingway, the second wife of Ernest Hemingway, and Dwight D. Eisenhower.

Pauline Pfeiffer married Ernest Hemingway in 1927; their union produced two sons, Patrick and Gregory, before they divorced in 1940. Gregory had a turbulent life, even by Hemingway standards. In 1951, Pauline telephoned Ernest to tell him that Gregory had been arrested. The conversation devolved into angry accusations. Some hours later, Pauline developed severe abdominal pain. She died shortly afterward during emergency surgery. Ernest blamed Gregory and his troublesome behavior for his mother's death. Years later, Gregory attended medical school and read the report of his mother's autopsy: she had died of a pheochromocytoma. He believed that her final argument with his father set off a crisis in which the tumor spilled excessive amounts of adrenaline into her system, causing her death.

Dwight D. Eisenhower, the supreme commander of Allied forces in Europe during World War II and president of the United States from 1953 to 1961, had the first of at least six heart attacks in 1955, and died almost fourteen years later from severe heart failure. Eisenhower's autopsy at Walter Reed Army Medical Center revealed a major surprise: his left adrenal gland harbored a pheochromocytoma. Eisenhower had had hypertension for many years and apparently had headaches associated with increases in blood pressure. In the last years of his life, Eisenhower had spikes in blood pressure up to 200/120 mm Hg. These changes in blood pressure suggest that his pheochromocytoma had been active, releasing adrenaline that raised blood pressure and could have further damaged Eisenhower's heart, possibly contributing to his ultimately fatal heart failure.

Octopus Pot Syndrome

An unusual syndrome increasingly recognized over the past twenty years mimics a typical heart attack due to coronary artery disease. Patients with this syndrome experience sudden chest pain, often after stressful emotional events (such as the death of a close relative, fierce arguments, public speaking, or being the victim of a robbery), but further tests show that they do not have coronary artery disease.

Cardiac imaging reveals that their hearts have an unusual and characteristic appearance, ballooning outward due to impaired muscle contraction. This syndrome, initially described in Japan but now known to occur worldwide, is called takotsubo cardiomyopathy because the shape of the affected hearts resembles a traditionally shaped Japanese ceramic jar used to trap octopus—a *tako tsubo.*[9] Takotsubo cardiomyopathy can be fatal; however, most patients fully recover and only infrequently suffer a recurrence. The exact cause of takotsubo cardiomyopathy remains uncertain. However, heart damage mediated by very high concentrations of adrenaline and noradrenaline induced by emotional stress is a leading potential explanation. Illicit use of cocaine or amphetamines, drugs that amplify the effects of adrenaline and noradrenaline, can cause similar damage to heart muscle.

Frightened to Death

In 2009, the *New York Times* noted the passing of William Devereux Zantzinger, who in 1963 went to jail for his role in the death of Hattie Carroll. Zantzinger, a young white tobacco farmer, had been drinking before arriving at a hotel in Baltimore for a formal-dress ball. Wearing a top hat and carrying a toy cane, he continued to drink heavily, struck a number of employees with his cane, and made offensive remarks about African Americans. He ordered a drink from Carroll, a fifty-one-year-old African American waitress and the mother of eleven children, but soon lashed out verbally and then struck her with his cane. Carroll retreated to another part of the hotel, telling co-workers that she felt very ill, and she died several hours later in the hospital from a stroke. Zantzinger was initially charged with disorderly conduct, but her death elevated the charge to murder.

Zantzinger's trial addressed how his actions contributed to Carroll's death. She had a history of hypertension and heart disease, and died from a stroke caused by bleeding into her brain—a cerebral hemorrhage. Hypertension is a prime risk factor for hemorrhagic stroke. An expert witness for the prosecution suggested that Zantzinger's verbal and physical attack on Carroll raised her blood pressure, which precipitated the stroke. The court concluded that the "combination of words and blow . . . caused the blood pressure to rise, resulting in the cerebral hemorrhage that caused her death," and ruled that Carroll's predisposition to a stroke on account of hypertension did not cancel Zantzinger's responsibility for his actions.

Carroll's story came to the attention of the young Bob Dylan, who was just beginning his eventually legendary career. Dylan wrote and recorded the song "The Lonesome Death of Hattie Carroll" soon after the trial. The lyrics of Dylan's song seem to suggest that the blow from Zantzinger's cane was what killed Carroll, but in fact adrenaline and the closely related substance noradrenaline almost certainly played a significant role in her death, a consequence of the actions of the sympathetic nervous system.

Threats perceived by the brain activate the body's sympathetic nervous system, which penetrates into the heart, blood vessels, and other major organs. The sympathetic nerves release noradrenaline, a slightly simplified version of adrenaline, which activates cells in these tissues. The sympathetic nervous system is vitally important in regulating physiological functions that are not generally under voluntary control, such as heart rate and blood flow. These nerves also connect with chromaffin cells in the adrenal medulla that synthesize and secrete adrenaline into the blood, where it is transported to distant organs. The functions of the adrenal medulla and many sympathetic nerves are complementary and often are activated together. They prove their worth in response to stressful stimuli.

When a person is, say, running from a ferocious dog, the activation of the sympathetic nervous system changes and integrates the function of organs in favorable ways. It increases the output of the heart in order to pump more oxygen-rich blood full of nutrients to the rest of the body; increases blood flow to the muscles and away from other organs where it is not immediately needed, such as the intestines; opens the lungs to breathe in more oxygen; and cuts blood flow to the skin to limit bleeding in case of injury. In the 1920s, Walter Cannon at Harvard Medical School developed a conceptual framework for the integrative capacity of the autonomic nervous system, especially in response to stress, labeling it the "fight-or-flight" response.

Blood pressure is not static but rather increases and decreases over the course of the day. Many common activities, such as mental and physical exercise, sexual activity, and caring for children, can substantially raise blood pressure. A primary mechanism for sudden increases in blood pressure involves the release of adrenaline and noradrenaline due to stress. Adrenaline and noradrenaline raise blood pressure by constricting arteries and increasing the output of blood pumped by the heart. In healthy people, the arteries in the brain readily tolerate these transient elevations of blood pressure. However, patients with long-standing hypertension, especially if not adequately treated, may develop progressive damage to these brain arteries. Over time, these pathological changes predispose these small arteries to burst, especially when blood pressure rises significantly in response to stress. Hattie Carroll had an increased risk for a cerebral hemorrhage due to these chronic changes; Zantzinger's actions that night likely drove up her blood pressure, putting into motion a series of physiological events that culminated in her fatal stroke.

Perils of an Insufficient Sympathetic Nervous System

Until this point, we have been discussing the adverse effects of excessive adrenaline and noradrenaline. On the other hand, too little of these substances can also severely interfere with health. An isolated deficiency of adrenaline alone has subtle effects. Genetically altered mice that cannot synthesize adrenaline look quite well resting in their cages. However, the stress-induced responses of their cardiovascular system are impaired. These results point to the value of adrenaline in integrating our capacity to deal effectively with physical challenges. However, a deficiency in the function of the sympathetic nervous system has far more obvious adverse consequences.

In 1925, Samuel Bradbury and Cary Eggleston described three patients who felt light-headed or fainted due to a marked fall in blood pressure when they stood up (postural hypotension). The sympathetic nervous system normally prevents gravity-induced pooling of blood in our legs on standing by contracting smooth muscle found within the walls of veins. Impaired sympathetic nervous system responses diminish the return of blood to the heart, leading to low blood pressure and fainting due to inadequate blood flow to the brain. Now known as the Bradbury-Eggleston syndrome, this type of postural hypotension is caused by impaired release of noradrenaline by the sympathetic nervous system.

Some astronauts returning from extended stays in low-gravity conditions in space also have low blood pressure on standing immediately after landing on Earth. This deficiency may be a case of "use it or lose it": in space there is no need for the sympathetic nervous system to fend off gravity-induced pooling of blood, and it takes a while for the sympathetic nervous system to readjust its activity after the return to Earth's gravity. Analogously, postural hypotension may develop in patients who are bedridden for weeks, especially the elderly. In addition, patients with diseases that can damage sympathetic nerve fibers, such as diabetes, can also have postural hypotension. Midodrine, an adrenaline-like drug that mimics some the effects of the sympathetic nervous system, helps raise blood pressure on standing in many of these patients.[10]

Challenging problems involving both underactivity and overactivity of the sympathetic nervous system may occur with damage to the spinal cord. A traumatic injury to the spinal cord in the neck can cut off the brain's capacity to regulate the sympathetic nervous system. Quite predictably, these people tend to have low blood pressure when upright. However, some patients may also have episodes of severe high blood pressure due to intermittent overactivity of the sympathetic nervous system (a condition called autonomic dysreflexia).

The recognition and understanding of autonomic dysreflexia occurred slowly, owing much to large centers that treated many wounded soldiers in wartime. During World War I, Henry Head and George Riddoch in England noted that some soldiers who had suffered an injury that severed their spinal cord had unusual symptoms associated with distention of their urinary bladders, including sweating, dilation of the pupils, and headaches. During World War II, Ludwig Guttmann, an expatriate German-born neurologist who established a national spinal injury center at Stoke Mandeville Hospital near London, observed the same phenomenon, and in 1946 he realized that blood pressure also increased to as high as 250/130 mm Hg during these episodes.[11]

We now understand the basis for these attacks of autonomic dysreflexia in patients with spinal cord damage. The spinal cord receives sensory inputs from neurons originating in organs throughout the body, including the bladder, bowel, and skin. For example, filling of the urinary bladder activates nerves that signal the spinal cord. In some people with a spinal cord injury, if the bladder is not soon emptied, these signals can trigger vigorous activation of the sympathetic nervous system, including a profound constriction of the blood vessels that raises blood pressure. This response does not occur in people with a normal spinal cord because the brain ordinarily suppresses this type of reflex response. However, after damage to the spinal cord in the neck, the brain loses its capacity to regulate the sympathetic neurons, whose activity is now uncensored. The best treatment is preventative: spinal cord injury patients learn to avoid situations that trigger these episodes. Occasionally, special treatment with antihypertensive drugs is required.

~:~

While in this chapter we have looked at some of the more dramatic effects of adrenaline and noradrenaline, these substances can have more subtle effects on the body as well. Chapter 2 will explore how science discovered that the body produces chemicals such as these, which regulate the function of other organs in myriad ways.

TWO

Ruled by Glands

What we feel and think and are is to a great extent determined
by the state of our ductless glands and viscera.

ALDOUS HUXLEY

By the nineteenth century, biomedical scientists generally believed
that the body's various functions were integrated by the brain alone,
through the signals sent along nerve pathways to the peripheral organs. This widely accepted paradigm delayed appreciation of hints
pointing to the existence of invisible chemicals released into the
blood that also influenced the function of distant organs. Now
we know that these chemicals, called hormones, secreted by endocrine glands, are vitally important. Just over a century ago, the putative existence of hormones remained shrouded in uncertainty and
controversy. We now live in a hormone-empowered culture, familiar
with the raging sex hormones of teenagers, the Pill, conundrums
surrounding female and male hormone replacement therapy, and
illicit use of growth and steroid hormones by athletes. The discovery of adrenaline in the adrenal glands marks a pivotal event in the
history of medicine: the breakthrough identification and isolation of
a hormone.

Two threads of scientific advance provide an important perspective on the discovery of adrenaline. The first thread involves the discovery of the adrenal glands in the sixteenth century and the emergence three hundred years later of evidence that these glands are
important in maintaining health. The second thread ties together
discoveries that pointed to the possibility that specialized glands

secrete substances into the blood that have activity elsewhere in the body. The convergence of these threads in the last decade of the nineteenth century culminated in the detection of adrenaline.

Discovery of the Adrenal Glands

Scientific knowledge of human anatomy took its first large strides more than two thousand years ago in Alexandria, Egypt, which at the time was a flourishing intellectual center that attracted numerous scientists. One of them was the Greek physician Herophilus, later known as the "father of anatomy," who made many important discoveries by systematically dissecting corpses (and possibly by vivisecting condemned prisoners), though he overlooked the adrenal glands. When Alexandria's influence faded, the scientific study of human anatomy markedly declined. Subsequently, in the second century AD, the study of animal anatomy reached an extraordinary peak with Galen of Pergamon, who early in his career was a surgeon to gladiators—one might say he was an early sports medicine specialist—and later served as physician to Emperor Marcus Aurelius in Rome. Galen made enormous contributions to research in anatomy and physiology by systematically dissecting a number of animal species, including primates, but he too did not describe the adrenal glands. For the next 1,500 years physicians revered Galen as the definitive authority in anatomy and medicine.

While medical science, along with many other branches of knowledge, went into a prolonged slumber during the Dark Ages in Europe, curiosity about human anatomy reemerged several hundred years later in conjunction with the desire of many Renaissance artists to accurately depict the human form. In 1543, Andreas Vesalius published *De humani corporis fabrica*, a comprehensive account of his exacting human dissections. This book transformed the science of anatomy with its accuracy; however, Vesalius also overlooked the adrenals. Finally, in 1563, Bartholomaeus Eustachius discovered the glands enveloped in fat near the top of each kidney. "I consider it indicated," he wrote, "to say something of the glands, diligently overlooked by other anatomists. Both kidneys are capped on the extremity . . . by a gland. . . . [E]arly anatomists and those who write ample treatises on this art in our days failed to detect them. They, pretending to be exact, stand so obstinately for their own and their masters' errors that often they seem to be fighters rather than searchers of anatomical truth."[1] These glands later became widely known as adrenal glands, from the Latin *ad*, "near," and *renal*, "kidney."

Eustachius's discovery received ruthless criticism from prominent physicians who retained reverence for Galen's ancient texts. Nonetheless, increasingly detailed information about adrenal anatomy gradually emerged, finally silencing the remaining dissenters. Cutting into an adrenal gland reveals two distinct parts that look different to the naked eye: an outer layer, called the cortex, and an inner layer, the medulla. Anatomists subsequently identified adrenal glands in mammals, fish, and reptiles; the widespread existence of the adrenals in many species spoke to their importance, but only in the softest whisper.[2]

Function of the Adrenal Glands

The function of some organs is immediately apparent from their anatomy: examples include the esophagus, the large tube that conveys food from the mouth to the stomach, and the major salivary glands, which have ducts that convey saliva into the mouth. On the other hand, the adrenals are solid organs without any ducts; their structure provides no hint about their function. For three hundred years following the discovery of the adrenals, many prominent physicians speculated about their function in the absence of evidence. Many believed that these glands primarily supported the nerves that crossed from the ribs into the abdomen. Others guessed that they were involved in urine production or sexual function. In 1716, the Académie des Sciences de Bordeaux in France organized a competition that asked the question "What is the use of the adrenal glands?" The youthful Charles de Montesquieu, who took an early interest in biology before becoming a great political essayist, judged the competition. One of the contestants asserted that the adrenals secreted a ferment into the kidneys through ducts whose existence was unknown; Montesquieu correctly commented that we would be perpetually ignorant of these putative ducts. Another contestant believed that the adrenals restored fluidity to the blood, and yet another that the adrenals filtered the blood. None of the submissions merited a prize. Montesquieu closed his deliberations by expressing the hope that one day the problem of the adrenals would be solved.

Later, in the eighteenth century, microscopic techniques provided more detailed structural information about the adrenals. The cells in the cortex look very different from the smaller cells in the medulla. Many nerve fibers terminate in the medulla. We now know that the secretion of steroid hormones from the adrenal cortex is controlled by chemicals in the blood, whereas adrenaline secretion from the medulla is activated by nerves. Nonetheless, these structural differences gave no indication on their own about the function or importance of the adrenals.

The next important insight came in the mid-nineteenth century from the observations of a physician based in a large London teaching hospital. Thomas Addison, the son of a grocer, received his medical degree in Edinburgh and after further training took up a position at Guy's Hospital, an institution with a culture that fostered medical discovery. Addison developed considerable standing as a diagnosti-

cian with major interests in correlating bedside findings in living patients with pathological changes found at autopsy. Addison consulted
on several patients who were experiencing severe fatigue along with
faintness, weight loss, vomiting, and mysterious darkening of the
skin; all these patients ultimately died. At autopsy, Addison noted
consistent damage in their adrenal glands. Through these observations, he described an unrecognized disease associated with destruction of the adrenal glands—organs with no known function. In his
book published in 1855, Addison modestly described his hypothesis
that diseased adrenals ("supra-renal capsules" in his terminology)
caused his patients' symptoms and deaths:

> It will hardly be disputed that at the present moment, the functions of
> the supra-renal capsules, and the influence they exercise in the general
> economy, are almost or altogether unknown. . . . To the physiologist and
> to the scientific anatomist, therefore, they continue to be objects of deep
> interest, and doubtless both the physiologist and the anatomist will be
> inclined to welcome, and regard with indulgence, the smallest contribu-
> tion calculated to open out any new source of inquiry respecting them.
> But if the obscurity, which at present so entirely conceals from us the
> uses of these organs, justify the feeblest attempt to add to our scanty
> stock of knowledge, it is not less true, on the other hand, that anyone
> presuming to make such an attempt, ought to take care that he do not,
> by hasty pretensions, or by partial and prejudiced observation, or by an
> over-statement of facts, incur just rebuke of those possessing a sounder
> and more dispassionate judgment than himself. Under the influence of
> these considerations I have for a considerable period withheld, and now
> venture to publish, the few facts bearing upon the subject that have
> fallen within my own knowledge; believing as I now do, that these con-
> curring facts, in relation to each other, are not merely casual coinci-
> dences, but are such as admit of a fair and logical inference—an inference,
> that where these concurrent facts are observed, we may pronounce with
> considerable confidence, the existence of diseased supra-renal capsules.[3]

Addison's discovery owed a debt to conceptual advances over the preceding 150 years in understanding the origins of disease. The key innovation involved comparing pathological findings at autopsy with the patient's symptoms in life.[4] Addison's discovery is breathtaking in its insightfulness, especially since he did not have the advantage of any modern laboratory tests. Addison did not know how destruction of the adrenals caused disease; one possibility he considered was that it involved damage to nearby nerves. The eminent French physician Armand Trousseau recognized the importance of Addison's conclusions and successfully advocated for the eponym "Addison's disease."[5] Addison lived long enough to see some confirmation and considerable criticism of this new disease. He had been plagued by depression for many years and committed suicide shortly after retiring in 1860.

Animal Models of Addison's Disease

If pathological destruction of the adrenal glands caused a fatal disease, then surgically removing the glands should lead to death. Charles-Edouard Brown-Séquard, an accomplished experimental neurologist, seized the opportunity to test Addison's conception.

Brown-Séquard's father, Charles Brown, was an American sea captain; his mother, Charlotte Séquard, was a woman of French descent who lived on Mauritius, an island in the Indian Ocean, once the habitat of the now extinct dodo bird.[6] Brown-Séquard grew up on the island with a passion for writing tales and plays. In 1838 he left for France, anticipating a career as a writer. However, a critic dashed his hopes, leading Brown-Séquard to choose medicine as a practical alternative.

Brown-Séquard's research style emphasized creativity and inspiration. He made enormous contributions to neurology, including elucidating the anatomical locations of specific neurological functions in the spinal cord.[7] In 1856, soon after learning about Addison's new disease, Brown-Séquard began studying the adrenal glands. He removed them from cats, dogs, rabbits, and guinea pigs and reported that the animals died within a day following surgery. However, his experiments soon became highly controversial.

One of those who took issue with Brown-Séquard's work was George Harley, who graduated in medicine from the University of Edinburgh in 1850. As a student he became locally famous for delivering a live, premature baby by Caesarean section immediately after the mother's death, single-handedly and without any prior experience. After graduation, Harley undertook extensive research training in outstanding laboratories in Europe. He returned to England as a physician associated with University College London. Realizing that speculative ideas about the adrenals had not yielded sound information, he started doing experiments to work out the function of the adrenals. Harley obtained inconclusive results with a limited number of adrenalectomies in cats and dogs, and so he moved on to experiments in rats. In 1857, Harley brought a healthy-looking white rat to a scientific meeting in London; he had removed its adrenals several months earlier, apparently without adverse consequences for the rat. The survival of rats after adrenalectomy convinced him that the adrenals had limited biological importance, contrary to the conclusions of both Addison and Brown-Séquard, and he directly challenged Addison at a meeting of the Royal Medical and Chirurgical Society in 1858. Harley's strongly expressed skepticism had considerable impact: many physicians remained unconvinced by Addison's concept and Brown-Séquard's results.

A major criticism of Brown-Séquard's experiments was that they did not exclude the possibility that the animals had died from complications of surgery, such as infection or bleeding, rather than specifically from the loss of their adrenal glands. In response to this criticism, Brown-Séquard did some mock operations that included all the manipulations short of actually removing the adrenals. These experiments did not provide compelling support for his conclusions since many of these animals also died, although not as quickly as in the original experiments. In addition, Brown-Séquard reported that some animals died after the removal of just one adrenal gland. With modern surgical techniques, removal of one adrenal is innocuous, as the remaining adrenal gland is sufficient to sustain life. Consequently, at least some of Brown-Séquard's experimental animals died from nonspecific surgical complications. Nonetheless, Brown-Séquard correctly concluded that the adrenals were essential for life, although the experimental basis for this conclusion and the biological explanation remained uncertain well into the twentieth century. On the other hand, Harley's focus on rats misled him about the adrenals. We now know that rats have tiny patches of adrenal cells elsewhere in the abdomen that can enlarge when the two adrenal glands are removed. Consequently, removing both adrenal glands in rats may be insufficient to kill them.[8]

Addison's disease provides a good example of the challenges faced in evaluating and integrating ideas that do not fit into an accepted conceptual framework. Addison made his discovery through careful observation and independent thinking. The controversies surrounding Brown-Séquard's experiments with surgical removal of the adrenals fogged the matter. In 1875, twenty years after the publication of Addison's book, Headlam Greenhow summarized what was then state-of-the-art knowledge about Addison's disease in a major lecture, noting that the medical community still did not generally recognize this new disease. The resolution of this controversy lay in the future, as it would require advances supporting the bold concept that glands secrete active substances into the blood.

The Birth of Endocrinology: A Very Long Gestation

Recognition that the ductless glands secrete active chemicals into the blood was the last major intellectual hurdle on the way to the discovery of adrenaline. Speculation about the potential role of secretions from glands had occasionally surfaced but led nowhere. For example, Théophile de Bordeu, an influential and flamboyant eighteenth-century French physician, is chiefly remembered for the partially correct guess that every organ—not only glands—secretes substances into the blood. However, his conjecture was not based on experimental data. His speculation without a substantive foundation represented the kind of sterile thinking Montesquieu had deplored earlier that century.

John Hunter in the eighteenth century and Arnold Adolf Berthold in the first half of the nineteenth century did interesting experimental work with roosters that came close to identifying hormones. Each of them made use of the fact that removing a rooster's testicles leads to the disappearance of the bird's observable male characteristics. Berthold's experiments suggested that reimplanting the testicles helped roosters retain these male features. However, neither Hunter nor Berthold conjectured that the testicles secrete into the blood a chemical that maintained maleness, and their results did not attract the interest of other investigators.[9]

Decades later, three lines of research provided an intellectual climate that led scientists to be much more inclined to accept the possibility that the adrenals secrete an active substance into the blood: a brilliant French scientist named Claude Bernard demonstrated that the liver secretes glucose, physicians and surgeons described a new disease involving the thyroid gland, and an elderly man claimed that he was rejuvenated by extracts made from sperm and testicles.

Bernard is another scientist who, like Brown-Séquard, yearned to be a writer but, not finding any encouragement from critics, took up medicine instead. It turned out to be a good choice, as Bernard devoted his career to strengthening the scientific foundations of medicine and made enormous contributions to physiology. His approach to science emphasized thoughtful, carefully controlled experiments.[10] Early in his research career, Bernard demonstrated that animals synthesized glucose. His elegant experiments in the 1850s established that the liver stored glucose as glycogen. We now know that glycogen is a starch, a huge molecule with thousands of glucose molecules strung together in chains with many branches. Bernard discovered that glycogen stores in the liver could be broken down to glucose, and that the freed glucose could be secreted into the bloodstream. This discovery represented an exciting, newly revealed interaction between organs: the liver influences other organs by providing them with glucose, a source of energy. The liver was well known to secrete bile into the gallbladder via a special duct; Bernard termed secretions into ducts *external secretions*. By contrast, he called the addition of a substance directly into the blood an *internal secretion*. These highly original results and concepts were published at about the same time as Addison's book on the adrenals.

The second line of research involved the thyroid gland, located at the front of the lower neck. Speculation about its function ranged from lubricating the voice box to heating the blood and even beautifying the neck. Progress in understanding the true role of the thyroid came very slowly. In the sixteenth century Paracelsus pointed to the potential clinical significance of swelling of the thyroid (goiter) by noting an association between goiters and cretinism. Cretinism is a congenital disorder characterized by stunted growth and mental disability that we now know is caused by low concentrations of thyroid hormone in the blood during infancy. Paracelsus's observation had limited impact because some individuals affected by cretinism did not have a goiter and most people with a goiter did not suffer from cretinism. As the second half of the nineteenth century began, physicians generally did not believe that the thyroid had much importance. Consequently, surgeons freely excised the entire thyroid gland to improve breathing in patients with very large goiters that compressed the main airway. Much surgical experience along these lines was gained in Switzerland, a country where widespread iodine deficiency caused many large goiters.

In 1859, Moritz Schiff showed that surgically removing thyroids from dogs and guinea pigs caused death. Schiff's work may have escaped general notice because he communicated his findings to a relatively small Danish research society and then buried the results in an article whose title suggested a different subject. At the time Schiff may not have appreciated the possible implications of his own results for patients who had undergone total thyroidectomies, but when speculation about internal secretions from the thyroid became a hot topic decades later, Schiff republished the work with additional experiments showing that reimplanting excised thyroids from the neck into the abdomen kept the animals alive.

In 1873, William Gull described a new clinical syndrome that ultimately would play an important role in connecting the thyroid to disease. Gull—a former student of Addison's—prospered as a physician who cared for members of the royal family. In a paper entitled "On a Cretinoid State Supervening in Adult Life in Women" he described previously normal women who developed physical and mental similarities to cretins. However, without being able to examine pathological samples of their thyroid glands, he had no way of knowing the condition of the glands. Still, he was one of the most famous physicians in England, and his report attracted considerable attention.[11]

In 1878, William Ord published several new cases similar to those described by Gull. Ord introduced a new term for the unusual swelling of the soft tissues of the face and limbs found in this disease, calling it *myxedema*. In one fatal case, the thyroid gland looked very abnormal at autopsy. Soon physicians throughout Europe and the United States began diagnosing myxedema, a new disease with an unknown cause.

At about this time, surgeons had increasing success removing thyroid glands. The Swiss surgeon Theodor Kocher had a very low operative mortality rate for thyroidectomies because of advances in general anesthesia, control of bleeding, and antisepsis during surgery. He did not initially pay much attention to the possible long-term consequences of the operation. In 1882, the surgeon Jacques-Louis Reverdin and his cousin noted that several of their thyroidectomy patients later became "cretins." Kocher quickly did a follow-up study of his patients and confirmed that total thyroidectomies had devastating long-term complications in some patients. As a consequence, Kocher advocated only partial rather than total removal of the thyroid.[12]

Felix Semon, a German laryngologist practicing in England who treated, among others, Queen Victoria (who knighted him) and King Edward VII, provided an important bridge between the worlds of physicians and surgeons. Semon knew about myxedema, and he soon learned about the new findings of Kocher and Reverdin. Brilliantly putting together what had appeared to be apparently separate diseases, he proposed at a meeting of the Clinical Society of London that cretinism, myxedema, and the complications from total thyroidectomy had a common origin: lost function of the thyroid gland. This bold assertion was initially greeted with polite skepticism. However, the Clinical Society later formed a committee to evaluate his contention.

Victor Horsley, a prominent member of that committee, surgically removed thyroid glands from a variety of experimental animals and noted that some of the animals developed symptoms of myxedema, although the results were inconsistent. Nonetheless, on the weight of the evidence, Horsley concluded that his results confirmed Semon's conjecture. In other words, failure of the thyroid, whether due to a disease of the thyroid or precipitated by surgically removing the gland, caused myxedema. In 1888, despite some confusing data, the committee endorsed this conclusion. However, its long and detailed report offered nothing substantial about treating the disease.

In 1891, a twenty-six-year-old physician named George Murray, a former student of Horsley's, grasped the possibility that the thyroid might produce a substance that prevents myxedema and that such a putative substance might be replaceable by an extract of a healthy gland. He presented his ideas at a local medical society meeting, only to hear a senior member of the society tell him that his proposal made no more sense than feeding an extract of spinal cord to someone with difficulty walking. Undeterred, and with strong encouragement from Horsley, Murray moved forward. He prepared a simple thyroid extract by grinding a fresh sheep thyroid in weak acid with some added glycerin and filtering this material through a handkerchief. Over the course of three months he repeatedly injected this extract into a forty-six-year-old female patient with severe myxedema. Remarkably, her symptoms progressively resolved. She would go on to live another twenty-eight years, over time receiving a total dose estimated to be the equivalent of 870 sheep thyroids.[13]

Based on this single patient, Murray cautiously concluded that the therapy really worked and did not represent a placebo response. He encouraged other physicians to test such extracts in patients with myxedema. Very quickly other physicians confirmed the efficacy of using thyroid preparations. William Osler, arguably the most well-known physician in the world at the time, later wrote about the benefits of this treatment in hypothyroid young children: "Not the magic wand of Prospero or the brave kiss of the daughter of Hippocrates ever effected such a change."[14]

In 1889, Brown-Séquard, now seventy-two years old and a very distinguished senior neurologist, made a scintillating announcement: he had been rejuvenated by injections of extracts of animal sperm and testicles. He did the experiment on himself as an outgrowth of a conjecture made twenty years earlier, when he proposed that debility in older men arose in part from diminished activity of the testicles. Brown-Séquard freely supplied extracts to other physicians in hopes of gaining independent evaluations.

Brown-Séquard's results created a sensation and raised a variety of objections: scientists rightly noted that the data supporting efficacy was minimal, antivivisectionists deplored the use of animal donors for this purpose, and social critics worried that efficacious treatment might potentiate masturbation. Brown-Séquard himself became concerned when the hyperbole surrounding testicular extracts reached the level of claims that they were an "elixir of life." Nonetheless, Brown-Séquard became increasingly enthusiastic about the efficacy of gonadal extracts. Modern methods for measuring the male sex hormone testosterone suggest that Brown-Séquard's extracts likely contained minuscule quantities of testosterone, much less than needed for pharmacological effectiveness. Consequently, the results he and others obtained probably represented placebo responses. Nevertheless, his intuition was again right even though his experimental foundation was weak. His claims stimulated widespread excitement about the possibility that extracts of glands might have major physiological effects.

In 1893, many physicians still believed Addison's disease only incidentally involved the adrenal glands. However, Murray's treatment of a patient with myxedema and Brown-Séquard's provocative assertions focused considerable attention on the potential power of glandular extracts. These developments provided additional motivation to look for active substances in the adrenal glands.

THREE

A Country Doctor's Remarkable Discovery

Till I heard Chapman speak out loud and bold:
Then felt I like some watcher of the skies
When a new planet swims into his ken . . .

JOHN KEATS

In the early 1890s, numerous medical journals reported both the benefits of thyroid extracts in myxedema and Brown-Séquard's rejuvenation claims. Then, as now, many physicians were enthusiastic— and sometimes uncritical—early adopters of novel therapeutic strategies, and so the use of animal extracts in patients expanded considerably. This prompted an editorial in the influential *British Medical Journal* in 1893 warning about a "possible epidemic of universal injections . . . often upon no better evidence than quacks produce for their 'cures.'"

At about the same time, however, George Oliver, a thoughtful physician in private practice, was conducting bold experiments with animal tissue extracts in healthy people. This chapter describes Oliver's seminal initial experiments and his follow-up collaborative work that led to the discovery of the substance in the adrenal glands that was later named adrenaline. The scientific and ethical aspects of his initial experiments are addressed from a modern-day perspective. The chapter concludes with an account of how Oliver's work had a major impact on the acceptance of the hypothesis that glands secrete active substances into the blood.

George Oliver's Provocative Experiment

In 1865, George Oliver graduated in medicine from University College London, where he learned from Professor William Sharpey about the importance of careful clinical measurements. Oliver started a general practice and later moved to Harrogate, a North Yorkshire town famous for its healing waters, where he became a respected consultant physician. During the tourist season, the medical problems of the many visitors in Harrogate kept him busy. With more leisure in the off-season, he tackled complex projects that satisfied his intellectual curiosity. His investigations ranged from urine testing to evaluating the composition and therapeutic value of Harrogate's waters.[1] He designed several novel laboratory instruments, including a mechanical arteriometer intended to measure the diameter of the radial artery at the wrist.

Oliver used the arteriometer to assess the effects of disease and drugs on arterial diameter. In 1893, he orally administered extracts of animals' adrenal glands, brains, and thyroids to several human volunteers. (Oliver obtained some of the extracts from the wholesale druggists Willows, Francis and Butler in London; the sheep and calf adrenal extracts likely came from a local butcher.) He concluded that brain and thyroid extracts caused the radial artery to dilate, whereas the adrenal extracts caused constriction. Had Oliver simply published these provocative but scientifically weak findings, they might have been ignored. However, he made the astute decision to seek collaboration with a dedicated scientist to extend these preliminary results.

Oliver introduced himself to Edward Schäfer, professor of physiology at University College London. They found some immediate common ground, as Schäfer too had been a student of William Sharpey.[2] During the winter of 1893–1894, Oliver and Schäfer intravenously injected adrenal extracts into anesthetized dogs. Their first experiment demonstrated astonishingly large increases in a dog's blood pressure within seconds of a single injection; this result provided the first compelling evidence of a specific physiological action of a glandular extract. The response lasted only a few minutes, but the effects of the extract could be prolonged by giving a continuous infusion. Oliver and Schäfer correctly inferred that the unknown substance in the extract was rapidly eliminated from the blood. They soon established that the active material resided in the adrenal medulla rather than the adrenal cortex.

Oliver and Schäfer next demonstrated that the adrenal extracts constricted arteries and dramatically increased the rate and force of the heart's contractions. Their experiments established that the extract acted on arteries directly, independent of the nervous system.[3] They made no effort to name the active substance that raised blood pressure.

A year later, working separately from Oliver and Schäfer, Wladyslaw Szymonowicz and Napoleon Cybulski at Jagiellonian University in Kraków, Poland, published substantially similar results from experiments with dogs.[4] They additionally found that the venous blood exiting the adrenals had the capacity to raise blood pressure, strongly suggesting that the adrenals secreted the active substance into the blood. Together, the British and Polish results provided provocative evidence that an unknown chemical in the adrenal medulla had dramatic effects on the cardiovascular system.

Oliver's Human Experiments: View through a Retrospectoscope

More than a hundred years later, a closer look at Oliver's human experiments illustrates the many changes in clinical investigation that have taken place since his day. His boldness in embarking on experiments with adrenal extracts still sparkles. However, research in human subjects is now conducted with much greater methodological rigor. Indeed, there are important indications that Oliver's conclusions based on measurements with his homemade arteriometer are unsound.

Oliver never demonstrated that the results of the human experiments were accurate or reproducible. He did very few human experiments, all conducted without controls; the results may have been due to chance. We now know that adrenaline is poorly absorbed after oral administration; consequently, not much, if any, of the adrenaline in the extracts Oliver used would have reached his human subjects' radial arteries. Additional reservations about the results stem from the fact that other measurements Oliver made with the arteriometer either are physiologically unlikely or are at variance with accurate modern methods that reliably measure arterial diameter. On the other hand, the experiments in dogs using intravenous injections in Schäfer's laboratory had a very strong foundation in the accuracy of measuring blood pressure, and the results are entirely consistent with those obtained through modern methods.

Oliver's limited written descriptions of his momentous experiments left room for various reconstructions by several famous contemporaries. Henry Dale—a major figure we will meet in Chapter 6—heard stories ten years later about Oliver's initial experiments from people who had worked in Schäfer's laboratory at the time:

> Dr. George Oliver, a physician of Harrogate, employed his winter leisure in experiments on his family, using apparatus of his own devising for clinical measurements. In one such experiment he was applying an instrument for measuring the thickness of the radial artery; and, having given his young son, who deserves a special memorial, an injection of an extract of the suprarenal gland, prepared from material supplied by the

local butcher, Oliver thought that he detected a contraction or, according to some who have transmitted the story, an expansion of the radial artery. [In his book *Pulse-Gauging*, Oliver states that the oral administration caused constriction.] Whichever it was, he went up to London to tell Professor Schäfer what he thought he had observed, and found him engaged in an experiment in which the blood pressure of a dog was being recorded; found him, not unnaturally, incredulous about Oliver's story and very impatient at the interruption. But Oliver was in no hurry, and urged only that a dose of his suprarenal extract, which he produced from his pocket, should be injected into a vein when Schäfer's own experiment was finished. And so, just to convince Oliver that it was all nonsense, Schäfer gave the injection, and then stood amazed to see the mercury mounting in the arterial manometer till the recording float was lifted almost out of the distal limb.[5]

Thomas Elliott, professor of medicine at University College in London—another major figure we will meet in Chapter 6—also provided some commentary on Oliver's human experiments. He believed that Oliver's son had been given an adrenal gland extract by mouth that made him feel sick and caused constriction of the radial artery. Elliott was impressed that Oliver had not attributed the response to nonspecific effects but rather pursued the clue by taking the information to Schäfer in London.

Schäfer himself described George Oliver's initial visit to his laboratory without drama:

> In the autumn of 1893 there called upon me in my laboratory in University College a gentleman who was personally unknown to me, but with whom I had a common bond of interest—seeing that we both had been pupils of Sharpey, whose chair I at that time had the honour to occupy. I found that my visitor was Dr. George Oliver, already distinguished not only as a specialist in his particular branch of medical practice, and also for his clinical applications of physiological methods. Dr. Oliver was

desirous of discussing with me the results which he had been obtaining from the exhibition by the mouth of extracts from certain animal tissues. . . . Dr. Oliver ascertained, or believed he had ascertained . . . that glycerin extracts of some organs produce diminution in caliber of the arteries. . . . Although these conclusions were interesting, it was easy to see that results which were obtained under mechanical conditions which were somewhat complex and not easy of interpretation, could not be expected to decide an important physiological question of this nature, and that it was essential, in order to obtain exact knowledge of the action, if any, of such extracts, to conduct the investigations with the employment of all means at the disposal of the modern physiologist.[6]

No matter the validity of Oliver's human experiments, they had enormous heuristic value, drawing him to definitive animal experiments in Schäfer's laboratory. Ironically, Oliver's inconclusive experiments in humans have been used in the twenty-first century as a putative example of how human research does not require confirmation in animals. In *Sacred Cows and Golden Geese*, Ray and Jean Greek state that Oliver and Schäfer's animal experimentation only reproduced data already known. In fact, the animal experiments launched an era in biomedical research that had major implications for human health.

George Oliver's experiments also illustrate the vast changes in ethical standards involving human research that have occurred over the past hundred years. He held sole responsibility for the decision to initiate his experiments. Today, investigators planning human research must pass the scrutiny of independent institutional review boards designed to protect the safety of the subjects. Ten years before Oliver's work on adrenal extracts, Italian investigators had published work demonstrating that injections of an adrenal extract killed rabbits. Today, even the most lax review board would not rubber-stamp a proposal to give normal volunteers substances that are known to kill animals. Oliver did not have to obtain informed consent from volunteers, nor did he have to deal with government agencies concerned with safe manufacturing practices for experimental substances. On the other hand, in his day some investigators felt a duty to test new chemicals on themselves first, an activity not generally acceptable now.[7]

The Hormone Concept

In 1895, shortly after conducting experiments with Oliver, Schäfer reinvigorated the term *internal secretion* in an address to the British Medical Association. The term had been defined decades earlier by Claude Bernard from his research on the release of glucose from the liver into the blood. Schäfer emphasized evidence suggesting that the adrenal and thyroid glands, too, produced internal secretions—that is, secretions that go directly into the bloodstream rather than into a duct. And though he missed the opportunity to predict the existence of testosterone (he believed that the well-known consequences of removing the testicles resulted from surgically induced nerve damage rather than from a missing internal secretion), his work with adrenal extracts and later with pituitary extracts played a major role in launching the new discipline of endocrinology, the study of internal secretions from ductless glands.[8]

Charles Sherrington, one of the greatest neuroscientists of the twentieth century, elegantly described the impact of Oliver and Schäfer's work:

> Saturday, 10th March 1894, was a day of note for medicine and physiology. On the afternoon of that day George Oliver and Edward Schäfer demonstrated in the latter's laboratory to their fellow-members of the Physiological Society of that time the action of intravenous injection of a small dose of watery extract of the suprarenal capsule [adrenal gland].

I remember well, for I was among those present, the profound impression the demonstration made on the little meeting. It opened a scientific vista, a vista which increased for each of us as conversation proceeded. I doubt, however, whether any of us foresaw, even so, one tithe of the full harvest of knowledge which was to ripen and be garnered from that experiment. . . . I do not think that any of us, even seeing it, foresaw what, as we look back, seems now a self-evident inference, that that experiment would largely transform the physiology that was coming into a chemical science, a science whose transcendent interest lay in biochemistry.[9]

The conclusive demonstration of a physiologically active substance in the adrenal medulla fully justified the search for active substances in other organs. In fact, a major extension and clarification of ideas about internal secretions occurred at University College London only a few years later, when Ernest Starling, together with William Bayliss, discovered a novel substance released from the intestines into the blood that activated physiological responses in the pancreas.[10]

At the beginning of the twentieth century, Russia's Ivan Pavlov was the world's leading authority on the digestion of food. The breakdown of food is critically dependent on multiple enzymes that flow from the pancreas through a duct into the intestines. Pavlov believed that nerves coordinated the concurrent arrival of food and digestive enzymes in the intestines. However, in 1902, Bayliss and Starling severed the nerves to the pancreas and found that the presence of food in the intestines nonetheless stimulated secretion of these enzymes from the pancreas. They discovered that injecting an extract of the wall of the upper intestine into the blood stimulated the pancreas to secrete digestive enzymes. They had demonstrated a novel, elegant control mechanism: arrival of food from the stomach into the first part of the intestine stimulates the intestinal cells to release into the blood a substance that activates the secretion of pancreatic enzymes, which reach the intestines through the pancreatic duct. Bayliss and Starling named this internal secretion from the intestines *secretin*.[11]

In 1905, Starling introduced the neologism *hormone* (from a Greek word meaning "I excite") for a chemical messenger that is released by an organ into the blood to control other organs.[12] This concept sharpened the earlier concept of internal secretion by assigning a name to chemicals that controlled the function of other organs. Accordingly, glucose remained an internal secretion of the liver—a fuel rather than a hormone—whereas adrenaline and secretin are hormones. Starling predicted that physiologists would soon discover many endogenous chemicals that controlled organ function. Since the word *hormone* emerged in conjunction with the discovery of secretin, some consider secretin the "first hormone." However, adrenaline meets all the criteria and has priority in meriting this honor from many points of view.

Over the past century, dozens of hormones have been identified, often based on paradigms established by Brown-Séquard, Oliver and Schäfer, and Starling and Bayliss. The pituitary gland secretes hormones that regulate the function of the thyroid, adrenal cortex, and gonads. The main hormone of the thyroid is thyroxine. The principal hormones of the adrenal cortex are cortisol and aldosterone, and the gonads of men and women secrete testosterone and estrogens, respectively. All these accomplishments owe a great debt to the seminal discovery of a blood-pressure-raising substance in the adrenal medulla. As we will see in the next chapter, after this discovery a fast and furious competition ensued to purify the active substance from the adrenal medulla and solve its chemical structure.

FOUR

Finding a Needle in a Haystack

He shall separate them one from another,
as a shepherd divideth his sheep from the goats.

MATTHEW 25:32

The Oliver and Schäfer experiments pointed to the existence of a chemical in the adrenal medulla that markedly raised blood pressure. Unmasking its identity presented a challenging biomedical problem that attracted ambitious scientists. We now know that adrenaline constitutes only about 0.1 percent of the weight of the adrenal glands, a tiny fraction of the total material, which made identifying it something like looking for a needle in a haystack.

More than a century ago scientists had very few clues about substances in the adrenal medulla and only rudimentary methods for isolating chemicals. In 1856, Alfred Vulpian discovered that iron-based stains tinted the medulla green without coloring the cortex. He also demonstrated that blood leaving the gland could be stained green, while blood entering the gland would not take stain. Since the stain did not color other organs, Vulpian concluded that the adrenal medulla released a distinctive substance into the blood. In 1885, Krukenberg demonstrated that a small molecule named catechol stained similarly to the adrenal medulla; he suggested that the stained material in the adrenal medulla might be related to a catechol.

This chapter describes the fiercely competitive race to purify adrenaline. The story includes a discussion of the two main and compelling protagonists—John Jacob Abel and Jokichi Takamine—who exemplify the complexity of assigning appropriate scientific credit for major discoveries and illustrate some of the different satisfactions of academic and industrial careers. The chapter addresses the interesting and important controversies surrounding the apparently simple assignment of a name to adrenaline and the validity of the patent related to this discovery. The chapter also describes a surprising use of royalties that flowed from adrenaline, which had an impact on Japanese-American relationships at the time and remains visible to this day.

Purification of Adrenaline

Schäfer asked Benjamin Moore, a junior scientist with a background in chemistry, to isolate the blood-pressure-raising substance in the adrenal medulla.[1] In 1895, after separating adrenal extracts into multiple fractions, Moore concluded that the substance stained with Vulpian's methods was the active ingredient that raised blood pressure. S. Fraenkel in Germany suggested that the active ingredient contained a catechol ring and a nitrogen atom, but he did not identify the full structure of the active substance. Moore determined that catechol itself was not the substance because it did not raise blood pressure. He also evaluated many chemical derivatives of catechol, but none of those raised blood pressure, either. Solving the problem would require an intense effort to purify the active substance to determine its structure.

In 1897, John Jacob Abel at Johns Hopkins University in Baltimore began purification efforts in his laboratory, starting with extracts from thousands of sheep adrenal glands. Using benzoyl chloride, Abel and his colleague Albert Crawford precipitated an amorphous yellow material that had some ability to raise blood pressure. They attempted to purify it further but unknowingly only partially succeeded in removing the added benzoyl groups that attached to the unknown active substance. In 1898, Abel reported progress in isolating the pure substance; his most purified material had the composition $C_{17}H_{15}NO_4$; that is, the substance contained atoms for carbon (C), hydrogen (H), nitrogen (N), and oxygen (O) in a ratio of 17:15:1:4. A year later he concluded that he had successfully purified the active chemical from the adrenals, and he named it *epinephrin*.[2] We now know that this assessment missed the mark; Abel had not removed one of the added benzoyl groups from the true natural substance.

Otto von Fürth in Strasbourg also tackled purification of the active substance from the adrenals. He got off on the wrong foot, proposing a structure that he soon abandoned. Subsequently, he used iron to precipitate a substance that he named *suprarenin*. Von Fürth believed that suprarenin was entirely different from Abel's epinephrin; moreover, he held that epinephrin had no connection with the active substance in the gland. Abel vigorously contested von Fürth's opinions and devoted considerable effort to sorting out discrepancies in the two accounts. However, while Abel was trying to refute von Fürth's claims, further up the East Coast a man named Jokichi Takamine was making better progress in isolating the active substance from the adrenal medulla.

Takamine had opened a small laboratory in New York City to isolate the blood pressure–raising substance in the adrenals. This work was stimulated by a suggestion from leadership at Parke-Davis, a pharmaceutical company based in Detroit.[3] Takamine hired Keizo Uenaka (sometimes written in English as Keizo Wooyenaka) from Japan, to do the experiments. Progress occurred very rapidly during the summer of 1900. Using large quantities of adrenal material supplied by Parke-Davis, Uenaka simply added ammonium to highly concentrated adrenal extracts. In a thrilling moment in July 1900, Uenaka saw crystals form in his test tubes. After dissolving the crystals, he evaluated their physiological activity using wild mice that he captured in the laboratory. Uenaka dropped some of the solution into the animals' eyes and watched with eager anticipation as the eyes turned white due to profound constriction of the arteries on the surface of the eyes. E. M. Houghton at Parke-Davis (who was an early contributor to the company's development of a diphtheria antitoxin) confirmed that this substance raised blood pressure.

Takamine named the new substance *adrenalin* at the suggestion of Norton Wilson, a friend and surgeon. Just a few months later, Thomas Aldrich, working independently at Parke-Davis, reported similar results.[4] Aldrich decided to follow Takamine's precedent and use the term *adrenalin* for his material. Aldrich's elemental analysis gave the formula for adrenaline as $C_9H_{13}NO_3$, whereas Takamine's material had a slightly different composition, $C_{10}H_{15}NO_3$. Aldrich's formula for adrenaline turned out to be the correct one, likely because his preparation had greater purity. Within months, industrial-scale production of Adrenalin (the name was registered as a trademark) began at Parke-Davis.

Abel initially believed that both Takamine's and Aldrich's chemical formulas were wrong. However, Abel soon realized that his own original formula was incorrect: he had not eliminated the last benzoyl group from epinephrin. Subtracting that group from his original formula left $C_{10}H_{11}NO_3$. He recognized that this was not quite right either, and he settled on $C_{10}H_{13}NO_3$. In 1905, his page-one article in the first issue of the now famous *Journal of Biological Chemistry* criticized the formulas of Aldrich and others while tenaciously supporting his own (incorrect) formula. However, other investigators, including von Fürth, Hermann Pauly, Gabriel Bertrand, Hooper Jowett, and George Barger with Arthur Ewins, confirmed Aldrich's formula. The story does have a curious twist: none of these scientists actually had pure adrenaline in their preparations. When the detailed structure of adrenaline became known, the elemental composition was indeed $C_9H_{13}NO_3$, as asserted by Aldrich. Nevertheless, another forty-five years would pass before the discovery that approximately one-third of the ostensibly pure adrenaline obtained from the adrenal medulla was actually noradrenaline, $C_8H_{11}NO_3$ (see Chapter 6).

The Chemical Structure of Adrenaline

The chemical structure of adrenaline—how the carbon, hydrogen, nitrogen, and oxygen atoms in the molecule are connected to each other—was determined rather quickly. Von Fürth and Pauly partially determined the structure; in 1904 Jowett narrowed down the possibilities to a few alternatives. Friedmann experimentally identified the correct option. Friedrich Stolz tackled the problem from the opposite direction, synthesizing the leading candidates for the structure of adrenaline in an attempt to find a molecule that fit adrenaline's biological profile. Stolz ultimately identified the correct compound in 1906 by virtue of its biological activity.[5] Henry Dakin independently synthesized the same compound shortly afterward in England.[6]

Unexpectedly, physiologists found that chemically synthesized adrenaline had only half the potency of adrenaline purified from adrenal glands; in other words, they needed to inject twice as much synthetic adrenaline as natural adrenaline to raise blood pressure by the same amount. A likely explanation involved the fact that synthetic adrenaline contained two mirror-image chemicals (isomers), analogous to left-handed and right-handed gloves. In 1908, Arthur Cushny demonstrated that adrenaline made in the body contained only the left-handed isomer (l-adrenaline). He inferred that right-handed adrenaline (d-adrenaline) had such little biological activity

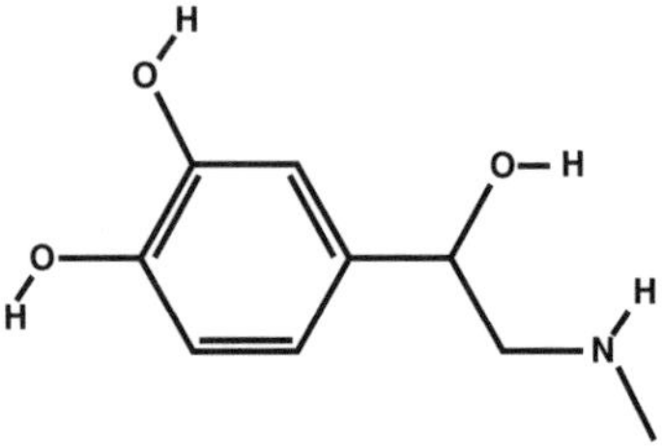

Adrenaline

Figure 4-1. The structure of adrenaline. Adrenaline is a small molecule. The benzene ring with the two attached hydroxyl (OH) groups is the catechol moiety (a functional group that is part of a molecule). The nitrogen (N) makes adrenaline an amine. Consequently, adrenaline and noradrenaline are often called *catecholamines*.

that only half of the chemically synthesized material is active. He confirmed this hypothesis by testing pure synthetic d-adrenaline obtained from Germany. We now know that the fit of adrenaline into adrenergic receptors is so tight and specific that only l-adrenaline activates these receptors.

For many years, companies sold adrenaline that either had been purified from animal glands or synthesized chemically. Purification of the natural substance involved costly steps since each adrenal gland contained only small amounts of adrenaline. Adrenal glands from about 10,000 cows might yield about 1 pound of adrenaline. Nevertheless, synthesizing large quantities of adrenaline was also challenging at that time.[7]

The Deeper Story: Abel, Takamine, and Adrenaline

Abel and Takamine had remarkably different life stories that converged for a time with their independent work on adrenaline. Looking a little more closely at their histories sheds light on how substances are named, how credit is assigned for scientific discoveries, and the ethics of patenting drugs.

John Jacob Abel

John Jacob Abel was born on a farm in Ohio in 1857, the oldest of eight children. He graduated from the University of Michigan and then studied at Johns Hopkins University in Baltimore before moving to Carl Ludwig's laboratory in Germany. During his seven years in Europe, Abel did research with several eminent scientists and earned a medical degree.[8]

In 1891, Abel returned to the University of Michigan as the first professor of pharmacology in the United States. Two years later he received a recruitment letter from William Osler in Baltimore, asking, "On what terms could you be dislocated?" This opportunity at the new Johns Hopkins School of Medicine proved irresistible, and Abel moved there as professor of pharmacology.[9]

Abel initially tried to isolate the active substance from the thyroid but had little success. In 1895, he redirected his research to isolate the active substance from the adrenals. Over the next decade he made many important contributions to purifying the substance, although in the end he did not fully succeed. During the course of this work he lost an eye in a laboratory explosion.

As previously noted, Abel invested enormous effort in defending his conclusions about epinephrin, acknowledging only much later that the substance he had isolated was not the pure form of adrenaline. In 1927, he reflected on those earlier days in a lecture: "The efforts of years on my part in this once mysterious field of suprarenal, medullary biochemistry, marred by blunders as they were, eventuated, then, in the isolation of a hormone, not in the form of the freebase but in that of its monobenzoyl derivative." In his obituary of Abel, William de Bernier MacNider cited Abel's lecture and pointed out the hidden human cost of the adrenaline research: "the great disappointment at times almost of an incapacitating character experienced by Professor Abel in not having actually carried the epinephrine work to its final point of obtaining the crystals and determining their structural formula. The indomitable spirit of the man overcame this disappointment to go forward for years to a variety of important discoveries."[10]

After finishing his studies of the adrenals, Abel worked on the isolation of other hormones. He crystallized highly purified insulin just a few years after Banting and colleagues in Toronto had first isolated cruder preparations of the hormone. He also worked on toxins in poisonous mushrooms, including *Amanita phalloides*, popularly known as the "death cap." In 1912 he made the surprising discovery that the parotid glands of the tropical toad *Bufo agua* contain much adrenaline; Abel had noticed that metal scalpel blades caused the toad's secretions to turn green, the color he had seen when cutting through adrenal glands years earlier. Abel also invented a device that served as a precursor of hemodialysis machines that are used to treat patients with kidney failure.

Abel had an enormous impact on the development of pharmacology in the United States. He took a broad view of the discipline, writing, "Let one pharmacologist be more of a chemist, another more of a physiologist, another more of a clinician."[11] He played a major role in the founding of the American Society for Pharmacology and Experimental Therapeutics (ASPET) in 1908 and a year later helped found the *Journal of Pharmacology and Experimental Therapeutics*, which he edited for more than twenty years.[12] In 1925 Abel received the Research Corporation of America's first annual research prize for "having done more to promote human enjoyment of life than any other living American scientist." Special interest focused on Abel's contributions to chemical research on the "ductless glands." A condition of the award was that the scientists made their contribution to knowledge "freely and without the hope of reaping a monetary reward from their endeavors."[13]

Jokichi Takamine

Born in Japan in 1854, Jokichi Takamine grew up during the period of rapid modernization that ended that country's isolation from the West. His father, a well-known physician descended from many earlier physicians, had considerable knowledge of Western medicine and spoke Dutch. His mother's family produced sake (rice wine), Japan's traditional alcoholic beverage. When Takamine was eleven, his family sent him to Nagasaki to complete his schooling and to learn English. In 1869, Takamine started medical school in Osaka; however, he found chemistry much more interesting, soon transferring his studies to Tokyo, where he graduated at the top of his class in applied chemistry from the newly established engineering school. His university instruction was in English, and his facility in English had a major influence on his subsequent life. In 1879, Takamine accepted an invitation from the Japanese government to join a small, select group of graduates bound for Scotland and England as part of a four-year program designed to facilitate implementation of modern Western methods in Japan. While abroad, Takamine learned many advanced methods in industrial chemistry, including for brewing alcoholic beverages and for manufacturing fertilizers. Takamine returned to Japan convinced that modern technology and science could improve traditional Japanese industries. He worked for the government in promoting industrial development, focusing on the manufacture of paper, dyes, and sake.

In 1884, Takamine temporarily left Japan for the United States as an official representative to the New Orleans World Exposition. This modest assignment transformed his life. He rented a room in a French Quarter mansion owned by Eben and Mary Hitch and soon fell in love with his landlords' eighteen-year-old daughter, Caroline Fields Hitch. A romantic connection between a Japanese man and an American woman was distinctly unusual in those days; nonetheless, they became engaged before his stay in New Orleans ended in 1885. In 1887, after establishing himself financially in the Japanese patent office, Takamine returned to New Orleans for the wedding. The couple's honeymoon was atypical, involving visits to fertilizer companies and a pause in Washington, D.C., where Takamine studied patent law. The couple also visited Niagara Falls, although not entirely as conventional newlyweds—Takamine took considerable interest in electricity generated by water power. Perhaps his wife was reminded of Oscar Wilde's remark made five years earlier: "Every American bride is taken there, and the sight of the stupendous waterfall must be one of the earliest if not the keenest disappointments in American married life."

Back in Japan, Takamine and his bride soon produced two sons. Takamine was an energetic entrepreneur: in addition to his government job, Takamine helped found the Tokyo Artificial Fertilizer Company in 1887, applying knowledge from abroad in using phosphate rocks to manufacture artificial fertilizer.[14] Takamine left his government job the next year; however, his company was slow to make a profit, as conservative farmers only hesitantly switched from traditional natural fertilizers.

With a considerable capacity to work on multiple projects simultaneously, Takamine also established a small laboratory to study fermentation. Brewing alcohol from grains requires two major steps: the long starch molecules in the grain must be broken down into small sugar molecules; subsequently, these sugars are transformed into alcohol. Traditionally, the Japanese use *koji* to convert starch from rice into simple sugars. *Koji* contains many microorganisms; especially important is the fungus *Aspergillus oryzae*.[15] The fungus contributes an enzyme—called a diastase—that breaks down the starch.[16] In the West, brewers typically obtain a different diastase from sprouted barley (malt). In both the East and the West, yeasts are the microorganisms used to convert the liberated sugar molecules into ethanol.[17]

Small batches of *koji* are typically cultivated with steamed rice, which provides nutrients so that the microorganisms in the *koji* can multiple many times over. In place of rice, Takamine developed a method of expanding batches of *koji* with wheat bran. Compared to rice, wheat bran—a by-product in making processed flour—is inexpensive. Takamine believed his form of *koji* production was cheaper, more effective, and faster, and he decided to apply his method to the manufacture of whisky. Coincidentally, his wife was growing increasingly uncomfortable in Japan as an American woman not well liked by her in-laws. Therefore, with considerable encouragement and support from his wife's parents, Takamine and his young family moved to Illinois in 1890 to apply his novel methods to brewing whisky in the United States.

At a time when technology typically flowed from West to East, Takamine swam against the tide, a rare Asian bringing technology to the United States. Takamine established his methods in Illinois, in part with scientific assistants recruited from Japan. The Takamine Ferment Company made considerable progress implementing these methods over the next several years in Peoria, an area rich in distilleries due to the presence of nearby grain farms, access to ample supplies of coal to provide a source of heat, and good water. His most influential connection to brewers evolved through the Distillers and Cattle Feeders' Company, otherwise known as the Whisky Trust. "Cattle Feeders" and "Distillers" fit together because, after extracting alcohol, the residual corn mash provided food for cattle. The Trust acquired exclusive rights to Takamine's process with the hope of producing whisky more cheaply.[18]

Takamine successfully conducted large-scale tests suggesting that his methods decreased costs and increased the yield of alcohol per barrel of corn. However, traditional malt producers vigorously resisted this potentially disruptive competition, including by stirring up xenophobia.[19] A mysterious fire at Takamine's experimental brewery slowed progress. Just as a major distillery operation was about to implement Takamine's process, the Whiskey Trust collapsed in a scandal involving financing, stock sales, and its business practices.

This commercial failure left Takamine's family almost destitute. At about the same time, Takamine became severely ill. During his convalescence Takamine got the idea of using his diastase for human digestive ailments. Demonstrating remarkable resiliency, he reinvented himself as the creator of the first enzyme used as a drug. He received a patent for purifying fungal diastase—the first U.S. patent granted for a microbiological enzyme. The enzyme found application as a treatment for indigestion and flatulence, and Takamine obtained a consultant position with Parke-Davis in Detroit. In 1897, Parke-Davis began selling the enzyme for medical use under the brand name Taka-Diastase, though Takamine retained the rights to distribute the enzyme in Japan.[20] In about 1898 the now more financially secure Takamine family moved to New York City; he established the Takamine Laboratory in nearby Clifton, New Jersey, to produce diastase for Parke-Davis. The company expanded, ultimately becoming a major supplier of many purified commercial enzymes.[21]

Soon after moving to New York, Takamine began an entirely new project: purify the blood-pressure-raising substance from the adrenal medulla. Parke-Davis recognized the potential therapeutic value of identifying the blood-pressure-raising chemical in the adrenals. William Warren, the general manager of Parke-Davis, asked Takamine to work independently on the problem. Perhaps to diversify their investment, Thomas Aldrich was given the same assignment in the company's own laboratories.

Takamine had considerable expertise as an applied chemist but lacked experience purifying small molecules. He visited Abel's laboratory at Johns Hopkins sometime around 1900 to gain additional insights into the problem, according to Abel's recollection published in 1927. Takamine told Abel that he thought his purification scheme could be simplified. Vague rumors later circulated suggesting that Takamine had somehow taken unfair advantage of Abel, but based on Abel's published papers, both before and after 1900, there is no reason to believe that Takamine obtained any information from Abel that he exploited inappropriately.

Effective research assistants can make the difference between success and failure in a challenging project. For Takamine, Keizo Uenaka's superb research experience represented enormous value. In Japan, Uenaka had taken a two-year technical training course, working in the laboratory of Nagajosi Nagai, a legendary professor of chemistry and pharmacy at the University of Tokyo.[22] Subsequently, Uenaka did chemical analysis at the National Hygienic Laboratory.[23] He believed that the freer atmosphere in the United States provided more opportunities for advancement, so he boldly left Japan and sought out Takamine in New York. Uenaka had a strong foundation from his highly relevant research experience to take on the challenge of purifying adrenaline. He was at the forefront of young Japanese scientists who continue making remarkable contributions in research laboratories throughout North America and Europe.

After Uenaka produced adrenaline crystals, Takamine filed a patent in the United States in 1900. As an industrialist, Takamine did not publish progress reports like the academically oriented Abel; Takamine revealed nothing about his work on adrenaline until he submitted his ultimately successful patent application. Afterward, however, he actively publicized his results at medical meetings and in publications that gave Keizo Uenaka some credit for his work (although neither authorship nor major financial rewards).[24]

Takamine was well aware of the dispute between Abel and von Fürth. In a paper published in January 1902 in which Takamine announced that he had isolated, in an active form, the substance in the adrenals that raised blood pressure, he modestly stated that he had no interest in taking credit from earlier investigators, but since they were involved in a dispute—and neither had identified the active substance—he felt entitled to name it.

Takamine: Contributions to Diplomacy and Industry

Takamine's life moved in a new direction at the time of the 1904–1905 Russia-Japan war fought over competing imperial interests in Korea and Manchuria. Japan surprised many Western governments by defeating Russia in major battles. Takamine contributed to Japanese public relations efforts by alerting Americans to Japan's scientific accomplishments and helping to raise loans for his government.

Takamine's great wealth—derived from sales of Adrenalin and Taka-Diastase—along with political involvement in Japanese relations with the United States, made him very socially prominent. He lived with his family in a mansion on Riverside Drive in New York City and acquired the replica Kyoto palace exhibited at the World's Fair in Saint Louis in 1904; the structure, named Sho-Foo-Den (Pine-Maple-Hall), became his vacation home in Merriewold Park in New York.[25]

Takamine co-founded the Nippon Club in New York City in 1905 and the Japanese Association of New York in 1914. He achieved considerable recognition as a leading, unofficial representative of Japan in the United States. He became increasingly concerned about relations between Japan and the United States and about the living conditions of Japanese nationals in the United States. Takamine's ongoing efforts to foster better understanding between Japan and the United States contributed to a great success that has endured for one hundred years. Little known to millions of visitors who have attended the annual National Cherry Blossom Festival in Washington, D.C., is the role Takamine played in bringing these trees to the United States from Japan.[26]

Takamine invested in Japanese-based companies involved in the production of chemicals, fertilizers, and dyes, as well as pharmaceuticals. He became president of Sankyo, a company that had its origins selling Taka-Diastase in Japan and which would later become a very large health care company. With great foresight, Takamine recognized that many important business opportunities would involve scientifically based industry rather than mechanically based manufacturing. In 1913, he proposed that Japan establish a national research institute. His articulate recommendation planted a compelling seed: several years later the Japanese government established Riken, a research institute for chemistry and physics, which became a major player in Japanese scientific research.[27] Takamine can be seen as one of the original biotechnology entrepreneurs, well before the deluge of biotech start-ups in the 1980s, and the purification of adrenaline represented his most notable scientific achievement.[28]

Perspective on Takamine and Abel in the Discovery of Adrenaline

Isolating adrenaline posed many challenges, especially given the relatively crude chemical purification methods available at that time. However, highly motivated scientists, a sufficient supply of raw materials, and reliable assays for measuring purified fractions led relatively quickly to the first crystals of adrenaline. The isolation of adrenaline involved "normal science" in the sense in which Thomas Kuhn uses that concept: challenging work that did not require block-buster ideas.

Simultaneous discoveries are a fact of life for scientists. Disappointment, frustration, or anger may emerge when one researcher is "scooped," or beaten to publication. Sole credit may go to the ones who publish first, sometimes leading to bitter disputes about priority. Coming in second in the solution of a scientific problem with a unique answer is especially frustrating, as many scientists are highly competitive and being first has implications for continued research funding, academic advancement, and recognition. While intense competition may stimulate hard work, its negative consequences include diminished cooperation and collaboration, duplication of effort, undue stress, and even scientific misconduct.

In the case of adrenaline, Abel and von Fürth sharply debated each other's results. Takamine remained outside the fray, pointing out his debt to earlier work and coming through with highly purified, biologically active crystals. Abel did not purify adrenaline before Takamine. Nonetheless, Abel is often remembered as the discoverer of adrenaline; he received Nobel Prize nominations for this discovery, while Takamine was ignored. In the twenty-first century, Abel is remembered in the United States and around the world as a great pharmacologist; Takamine's reputation lives on mainly in Japan.[29]

Terminology: Adrenaline versus Epinephrine

Even the dullest drug generally has three major names: a chemical name, a generic name, and a trade name.[30] The least used drug name spells out, in chemical language, the composition of the molecule, generally according to the rules of the International Union of Pure and Applied Chemistry. For adrenaline, the chemical name is 4-[(1R)-1-hydroxy-2-(methylamino)ethyl]-benzene-1,2-diol and is hardly used.

The World Health Organization (WHO) assigns generic names, officially called international nonproprietary names. Countries are not obligated to adopt the WHO recommendations; for example, the official name in the United States (known as the United States Adopted Name) and that in Britain (the British Approved Name) may be different from each other and from the international nonproprietary name. Desirable generic names are relatively simple and clearly distinct from the names of other drugs to avoid physician and pharmacist confusion in preparing and filling prescriptions.

The third type of name is the trade name (brand or proprietary name), assigned by the drug company owning the drug. The trade name—generally euphonious and easy to remember—may be pivotal in successfully marketing the drug to physicians and patients. Drug companies do have some regulatory constraints in inventing proprietary names for their new drugs, however.

The naming of adrenaline involves a surprisingly complex story played out over more than a hundred years, beginning when Abel named the substance he isolated from the adrenal medulla *epinephrin*. (This, we now know, was not the natural molecule, but rather still contained an added benzoyl group.) It should be noted that several years earlier the Polish scientists Wladyslaw Szymonowicz and Napoleon Cybulski had named the blood-pressure-raising substance *nadnerczyna* (*nadnercze* is "adrenal" in Polish), but this name did not stick elsewhere.

How did Abel come up with *epinephrine*? Though Abel was a former Latin teacher, he agreed with the renowned anatomist and lecturer Josef Hyrtl, who proposed that the Greek-derived *epinephris* (*epi*, "close by," and *nephros*, "kidney") was the best name for the adrenal glands. Hyrtl preferred Greek origins, alluding to a quotation attributed to the French playwright Molière: "Parce qu'avec du grec on a toujours raison," or "Because with Greek one is always right." Ironically, *epinephris* never caught on widely as a synonym for the adrenal glands.

Von Fürth, as we have seen, named his substance *suprarenin*. He did not follow Molière's admonition to rely on Greek but rather used a word with a Latin base: *supra*, "above," and *ren*, "kidney." The term *suprarenal* had a long-standing, established usage; indeed, Thomas Addison's 1855 book used the phrase "supra-renal capsules" in its title. Von Fürth's substance was not the sought-after hormone; nonetheless, Suprarenin later became one of the proprietary trade names for adrenaline marketed in a number of European countries.

The name *adrenalin* appeared in a scientific journal for the first time in 1901, in a paper by Takamine. The origin of the word *adrenalin* is Latin (*ad* meaning "near"; i.e., the substance comes from glands near the kidneys). Norton Wilson, a well-known otolaryngologist, visited Takamine's laboratory in October 1900; after seeing the crystallized substance that had not yet been named, Wilson suggested that name. Parke-Davis sold its product under the trade name Adrenalin (capital *A* with no *e* at the end), and succeeded in obtaining a U.S. trademark. Quite quickly, both *adrenalin* and *adrenaline* began appearing in publications as scientists reported results of their experiments. Takamine himself used *adrenalin* in a presentation in England in 1901 and in a paper in the *Journal of the American Medical Association* in 1902, without any reference to Parke-Davis.

Henry Wellcome, the director of the large British pharmaceutical company Burroughs Wellcome, tried to block Henry Dale, then working in the company's research laboratories, from publishing a paper using the word *adrenaline*. Wellcome had great enthusiasm for protecting commercial rights and objected to using a word so close to

Adrenalin, Parke-Davis's trademark. While physiologists preferred *adrenaline*, Wellcome's chemists adopted *epinephrine*. After a hot debate within the company, Wellcome relented and allowed Dale to publish his paper with the contentious word *adrenaline*.[31] No significant repercussions followed from Parke-Davis, as the company had not registered the name in England. However, Parke-Davis did win a lawsuit in Belgium in 1904 over the name Adrenalin and threatened others. In any case, the word *adrenalin* soon became incorporated into the English language, along with words such as *yo-yo, zipper*, and *escalator*, also originally trademarks.

In 1907, Thomas Maben wrote an impassioned review of the disputed naming of the hormone secreted by the adrenal medulla. He maintained that since Abel's original epinephrine was not the right chemical, *adrenaline* was the appropriate name, as it was the first to refer to the correct substance. In 1911, the *Journal of the American Medical Association* published a lengthy editorial concluding that the correct scientific name for the active substance in the adrenal should be *epinephrin* (without the last *e*) and used all the time, except when referring to the product of a specific drug company.[32] The editorial emphasized the importance of Abel's contributions and noted that Abel had done his work for the public good, freely publishing his findings without seeking or receiving any personal financial reward. Parke-Davis, however, complained about the journal's substitution of *epinephrin* for *adrenalin* even when the authors used the company's product. Parke-Davis felt entitled to the credit after heavily investing in Takamine's research and paying him a royalty on all the adrenaline sold by the company. The journal acknowledged that *adrenalin* appeared extensively in Europe as a generic term yet remained adamant that the drug should be called *epinephrin*, at least in its pages.

Debate about the name of the active substance found in the adrenal medulla continues up to the present day. The WHO nonproprietary name and the Unites States Approved Name are both *epinephrine*. On the other hand, its British Approved Name is *adrenaline*, which is also the name used in some other countries. In 2006, in honor of Takamine, Japan decided to change its official name for the drug from *epinephrine* to *adrenaline*. There is considerable pressure in the United Kingdom to adopt *epinephrine* in light of the European Union's commitment to using WHO-recognized international nonproprietary names. However, physicians have raised the concern that changing well-entrenched drug names could lead to dangerous confusion in hospitals. A letter to the editor of the *British Medical Journal* in 2000 from a physician in Greece noted that while the medical term for the adrenal glands was *epinephridia* in Greek, he and his colleagues shouted for adrenaline in emergencies.[33]

Lawsuit about Substance: Adrenaline on Trial

A 1904 lawsuit escalated the force and consequences of arguments surrounding priority in the discovery of adrenaline. Parke-Davis's sales of Adrenalin were being threatened by competition from the H. K. Mulford Company in Philadelphia and other companies that launched purified adrenaline products. Lawyers representing Parke-Davis wrote to Mulford asserting that Mulford's product Adrin infringed on Takamine's patent for Adrenalin and asking the company to immediately stop this practice. Mulford replied that it had not infringed on any rights and therefore could not comply with Parke-Davis's request. Parke-Davis consequently launched a lawsuit to protect its rights under the patent. Establishing Takamine's position as the discoverer of adrenaline became an important legal issue.

In general terms, patent systems protect the rights of the inventor while simultaneously forcing public disclosure of key information about the invention. The disclosure requirement represents a fundamental trade-off: other scientists and inventors can use the information to make their own new discoveries but cannot copy the patented invention without the permission of the patent holder. The possibility of patent protection with its attendant time-limited monopoly on a product encourages companies to invest in research.

The authority for patent protection in the United States goes back to the Constitution, and the Patent Act of 1790. In 1800, a revision in the law allowed foreigners to hold U.S. patents—a right that helped Takamine a hundred years later. However, enthusiasm for patents has fluctuated over time. At the beginning of the twentieth century, university-based scientists generally did not patent their inventions; John Abel made no attempt to patent his work with epinephrine. Moreover, many people opposed patenting drugs. For a considerable time, the American Medical Association's Code of Ethics opposed allowing physicians to hold patents on medicines. The association later recognized that precluding patents on drugs would promote secrecy, with adverse consequences. Values have changed extensively since those days, however; in the twenty-first century, research-intensive universities devote considerable resources to exploiting patentable discoveries made by their faculty in the hope of making money as well as fostering commercialization of important innovations.

In his patent application, originally filed in November 1900, Takamine was represented by lawyers from the New York office of Knight Brothers, a highly experienced legal firm. Takamine's application aimed to patent not just the procedure used to isolate adrenaline but also adrenaline itself. If his application was successful, he would have the rights to adrenaline, no matter what purification method was used to obtain it. Takamine's patent application had a rigorous review by James Littlewood, a highly capable examiner who had a medical degree from Georgetown University. Littlewood quickly decided, based on established criteria from other cases, that because adrenaline was a naturally occurring substance, it could not be patented. Takamine's lawyers then split off the isolation procedure claims into a separate application, which readily moved forward. They also submitted a new application solely addressing the claim that adrenaline itself could be patented. After considerable back-and-forth with Littlewood, the new application was revised, claiming that purified adrenaline was not exactly the same substance found in the adrenal extracts; as an example of a difference, they indicated that it was much more chemically stable than adrenaline in a gland extract. After more than two years of skirmishes, Littlewood finally relented, allowing the claim for adrenaline as a modified or improved form of a natural substance.

In 1911, Judge Learned Hand of the Circuit Court of the Southern District of New York decided Parke-Davis's patent infringement lawsuit against the H. K. Mulford Company. While some of the testimony was nontechnical—for example, Frank Ryan, the president of Parke-Davis, testified about the relationship between his company and Takamine and indicated that Takamine received a 5 percent royalty on the wholesale price of Adrenalin—much of the testimony involved extended examinations, cross-examinations, and rebuttals from expert chemists who had studied all the relevant purification papers in extraordinary detail. Mulford claimed that Takamine was not the first to purify adrenaline and so could not patent it. Charles Chandler, a very famous industrial chemist, professor, and highly experienced expert witness in patent litigation, testified on behalf of Parke-Davis. Mulford relied on another major figure in American chemistry, Samuel Sadtler, as its expert. Questioning of each of these witnesses was exhaustive. Judge Hand spent considerable time and effort coming to grips with the voluminous and at times contradictory testimony of these formidable expert chemists. (Interestingly, a decade earlier Hand had written an article published in the *Harvard*

Law Review on problems associated with having paid expert witnesses battle it out in court.)

Judge Hand's decision in favor of Parke-Davis had considerable impact, both immediately and as a precedent. He upheld the patenting of the natural substance adrenaline, stating that there was no rule against its being patented even if unchanged from its natural state. This landmark decision served as a precedent many years later in decisions involving the patenting of genes.[34] However, his decision on the patentability of natural substances has been criticized and remains controversial a full century later.[35]

While the battles about nomenclature and patents were going on, purified adrenaline was almost instantaneously exploited by physicians to treat an amazingly wide range of disorders in clinics and hospitals around the world. These clinical applications and the lessons learned from widespread use of a drug before it has been extensively evaluated will be taken up in the next chapter.

[FIVE]

Adrenaline Zips from Bench to Bedside

Adrenaline can be made to revivify the heart of the dead child.

LOS ANGELES TIMES, OCTOBER 1, 1901

Adrenal extracts and then adrenaline moved rapidly from the laboratory to the bedside. The initial infectious enthusiasm for adrenaline contributed to physicians prescribing this new drug in highly fanciful and futile efforts to combat disease. Physicians learned only gradually and inconsistently from anecdotal experience about adrenaline's efficacy and serious adverse effects. This chapter describes both the rational and the indiscriminate incorporation of adrenaline into medical practice. These stories, carried through to the present, illustrate the advances that have been made over the past century in the systematic evaluation of the safety and efficacy of drugs, as well as limitations in this process.

Early Use of Adrenal Exacts and Adrenaline in Patients: The Good, the Bad, and the Ugly

Shortly after Oliver and Schäfer demonstrated that adrenal extracts constricted blood vessels, Oliver noticed that smearing the extracts on incisions in dogs stopped the bleeding from arteries cut by his scalpel. In 1896, the ophthalmologist William Horatio Bates reported that sheep adrenal extracts considerably improved conjunctivitis by constricting blood vessels in red eyes. He found that the extracts

66

decreased bleeding during eye surgery; he soon expressed great enthusiasm for using the extracts to stem the vigorous bleeding that occurred in nasal surgery. Bates reported hearing about cases where the extracts helped decrease stomach hemorrhages and bleeding from the uterus after childbirth. Bates also believed that adrenal extracts decreased bleeding in hemophiliacs.[1] He tried unsuccessfully to isolate the active substance in the extracts.

In a review published in the prestigious *Journal of the American Medical Association* in 1900, Bates made wide claims about the benefits of adrenal extracts, including that they could make a slow heartbeat faster and a fast heartbeat slower. Bates did not express reservations about these pharmacologically contradictory assertions. He believed that the clinical benefits were so obvious that a stronger scientific foundation was unnecessary. In the absence of adequate research methods, many inevitable errors littered the clinical landscape. For example, Bates thought that these extracts cured gonorrhea-induced eye infections and helped deaf people hear.[2]

As soon as Parke-Davis began marketing adrenaline, the purified drug quickly made major inroads on the use of adrenal extracts. The marketing of adrenaline by this American company represents a major milestone: a powerful drug born from research conducted in the United States and then sold in Europe and around the world. Fletcher Ingals published a paper about the clinical benefits of the drug in the April 27, 1901, issue of the *Journal of the American Medical Association*. He acknowledged receiving "adrenalin" from a prominent drug company, which he left unnamed. Ingals reported that his methodology involved experimenting with adrenaline in multiple cases and keeping careful notes. Without further ado, he concluded that adrenaline worked very well in several conditions, especially upper respiratory congestion. In 1903, Heinrich Braun, an eminent German surgeon and pioneer in local anesthesia, emphasized that adrenaline potentiated the beneficial effects of the local anesthetic cocaine in surgical procedures. By causing vasoconstriction, adrenaline not only diminishes bleeding but also slows down the rate at which cocaine is carried away from the surgical site by blood flow. In turn, cocaine potentiates the effects of adrenaline (discussed in Chapter 6). Braun's work with adrenaline stimulated wider use of combination therapies with cocaine and drugs related to novocaine. By 1905, Parke-Davis exploited these developments by marketing Eudrenine, a combination of adrenaline and the local anesthetic beta-eucaine; in addition, Parke-Davis combined adrenaline with cocaine.

In 1901, a pound of adrenaline reportedly cost \$7,000 (approximately \$180,000 in 2012 dollars). Sales of purified adrenaline grew enormously as physicians and surgeons quickly found uses for the drug within their areas of special expertise. A Parke-Davis manual recommended using a camel's-hair brush, a spray, or other suitable applicator to put adrenaline on the site of bleeding or inflammation. Potential therapeutic targets included the eyes, ears, nose, throat, vocal cords, airways, stomach, intestines, urethra, hemorrhoids, vagina, and uterus. Moreover, physicians found that adrenaline ameliorated the discomfort from runny noses in people with hay fever and cleared up hives in patients with allergies.[3] Adrenaline later emerged as the major treatment for people with anaphylaxis, a life-threatening allergic response characterized by difficulty breathing, hives, and low blood pressure.[4]

Physicians tried adrenaline in an astonishing range of diseases. Fanciful claims emerged that adrenaline stopped the growth of cancers in the mouth and dried out large fluid collections in the lungs and abdomen. Several physicians in India reported that adrenaline saved patients otherwise doomed by the bubonic plague—the Black Death.[5] Perhaps less dramatically, bed-wetting reportedly resolved with adrenaline, and patients with large, bulbous, red noses—rhinophyma—improved with adrenaline. Even symptoms of gonorrhea apparently melted away after irrigation of the urethra with a heated salt solution and urethral injections of adrenaline. The *New York Times* recounted a French physician's assertion that adrenaline injections into prospective parents just prior to conception determined their offspring's gender.

These claims typically involved reports by physicians who treated as few as one patient. The reports had major biases as both the physicians and patients had great expectations for this new drug. Conclusions about efficacy often depended on how patients felt before and after treatment rather than on objective measurements. Without comparing the effects of adrenaline to placebos, the extent of spontaneous improvement could not be estimated. In modern times, investigators design large clinical trials that are objective, statistically valid, and free from bias to rigorously test the efficacy of new drugs.

During a devastating polio epidemic in New York City in 1916, claims about the efficacy of adrenaline promoted much ill-founded optimism. Samuel Meltzer, a physician-scientist at the Rockefeller Institute (now Rockefeller University), recommended injecting adrenaline into the spinal fluid of children with acute polio several times each day.[6] He based the treatment on very limited experiments with adrenaline in virus-infected monkeys. A number of children received this treatment, at best without apparent harm. News reports exaggerated the outcome; for example, the *Atlanta Constitution* noted on July 28, 1916, "Intraspinal injections of adrenalin have proven effective in all 50 cases so treated." However, polio experts expressed considerable skepticism; the New York health commissioner weighed in that the treatment had not been adequately studied and that physicians who used it were doing so on their own responsibility. Experiments on polio patients continued into the 1930s, including risky injections of an adrenaline analog directly into the fluid at the base of the brain. No benefit was ever established.

The widespread use of adrenaline to curb inflammation in the eyes received considerable publicity. Immigrants to the United States with trachoma quickly took advantage of these results to pass on-the-spot medical examinations. At the turn of the century, trachoma was a largely untreatable, contagious eye disease that could progress to blindness. The return of Napoleon's army from Egypt, where the disease was endemic, contributed to the spread of trachoma in Europe. Having trachoma prevented a prospective immigrant from entering the United States.[7] Medical screening of immigrants depended heavily on rapid assessments. In some phases of the disease, the eyes appear red with inflammation. Given the high stakes, immigrants with trachoma tried hard to get past the inspectors. In 1905, the medical journal *Lancet* reported that officers in the U.S. Public Health Service at Ellis Island in New York had noticed that the eyes of some immigrants looked strangely blanched. In holding these people for reexamination, they noted several hours later that some of these individuals had obvious eye congestion indicative of trachoma. With these clues and even closer scrutiny, officers caught immigrants with trachoma instilling adrenaline into their eyes just prior to inspections. Adrenaline obliterated evidence of mild trachoma for about thirty minutes.[8]

Adrenaline soon moved into the boxing ring. Specialized experts leap into the ring between rounds in professional matches with sixty seconds to stop the bleeding from their fighter's wounds. In addition to pressing ice-cold, smooth metal endswells against wounds, cutmen use adrenaline-soaked cotton swabs to constrict bleeding arteries (now sometimes augmented with thrombin- or collagen-containing materials to accelerate blood clotting). The apparent benefit of adrenaline compared to its possible harm in boxers has not been evaluated in clinical trials, though absorption of adrenaline into the blood could have adverse effects on the heart and arteries in stressed boxers. As early as 1936, concern about the effects of adrenaline on cuts arose after the fight between the victorious former heavyweight champion Primo "Ambling Alp" Carnera and his opponent Isidoro "Izzy" Gastanaga. The fight ended in the fifth round due to a cut near Gastanaga's eye; the *Washington Post* reported that referee Arthur Donovan believed the eye was "damaged as much by the emergency use of adrenalin as it was by the cut."

Adrenaline: Shock and Awe

Adrenaline's capacity to raise blood pressure led naturally to its use in patients with low blood pressure. George Crile, part of whose training involved an extended visit in 1895 to University College Hospital in London, where he met Schäfer, pioneered early experimental work that tested the value of adrenaline in maintaining blood pressure during surgery. His physiologically based research had very high standards of scientific rigor, quite different from the haphazard clinical "experiments" with adrenaline done by many of his contemporaries.[9]

Crile had a deep interest in low blood pressure and shock, particularly in surgical patients. Shock is a life-threatening condition characterized by inadequate blood flow to organs, typically caused by blood loss, abnormal dilation of blood vessels, or damage to the heart. Crile tried many commonly used remedies in the treatment of shock during surgery but found them ineffective. He developed a pneumatic suit—a precursor to modern antigravity suits—to press blood from the legs back to the heart to combat shock. He tried infusing salt solutions to increase blood volume in an effort to raise blood pressure.

He concluded that during general anesthesia and surgery, the brain lost the capacity to control the tension in blood vessels; he realized that a drug that constricted arteries, building up the blood pressure, might be helpful. In a series of careful studies, he found that adrenaline provided the needed stimulus to maintain blood pressure.

In 1903, Crile summarized several years of animal research on physiological changes during surgery at a major meeting at Harvard Medical School. In his autobiography, Crile described the packed room: "The gray-haired, bald-headed, straight-backed shogun were all present and in the front pew, surrounded by their satellites in the order of precedence." He heard afterward that there were many side remarks during his presentation, such as "I don't believe a word of it" and "Impossible." Consequently, Crile was pleasantly surprised that his presentation actually stimulated plans to study his methods in major Boston hospitals.[10]

Crile turned his attention to the phenomenon of sudden death during surgery. He used excessive doses of anesthetics to stop the hearts of dogs, then tried many stimulants to revive the animals, but without any success. However, in the very first experiments with adrenaline, he exclaimed, it "gave us one of the greatest possible thrills." Crile administered adrenaline to a dog fully five minutes after its heart and breathing had stopped. "In ten seconds," he reported, "there was a tumultuous heart beat, an uprush of the blood pressure and a spontaneous respiration." His research group's excitement knew no bounds as they envisioned adrenaline reviving dead people: "A second chance at life might be provided." A reporter who had slipped into a scientific meeting and saw Crile resuscitate a dog spread the news worldwide, and Crile was inundated with letters, calls, and cablegrams, such as one from a widow who inquired about reviving her late husband, who had fallen years earlier into a crevasse and been enveloped by a glacier. Crile soon learned that resuscitation had to begin quickly, as a lack of blood flow damaged the brain in minutes.[11]

The news of Crile's work on reviving the dead with adrenaline had an enormous impact on popular views about the powers of modern medicine. For the next thirty years, numerous stories appeared in major newspapers describing with awe the capacity of adrenaline to revive the apparently dead. These cases ranged from babies born dead to drowning victims and people who had been lifeless for hours. In 1935, reports appeared indicating that a physician had frozen a rhesus monkey named Jekal for five days at 30 degrees below zero Fahrenheit and then revived the monkey with heat and injections of adrenaline. Relatively few articles expressed an appropriate skepticism about the results, however. (For more on popular views about adrenaline, see Chapter 10.)

Newspapers reported the use of adrenaline in treating famous people. President William McKinley was shot by Leon Czolgosz in Buffalo, New York, in September 1901, and required emergency surgery. The president initially did well but began to deteriorate a week later; when his pulse became rapid and weak, he received several injections of adrenaline, but he died nonetheless. Adrenaline was then so new that it was not explicitly named in newspaper reports, which indicated that the president had received "strychnine, digitalis, and other powerful heart stimulants." On the other hand, at the time of President Franklin Roosevelt's final illness in 1945, adrenaline was widely known; when Roosevelt had a cardiac arrest following a stroke, the *Washington Post* reported that his physician gave him an intracardiac injection of adrenaline in the vain hope of stimulating his heart into action. In between these presidential deaths, the *New York Times* reported in 1936 that Thomas Edison collapsed during the celebration of the fiftieth anniversary of the electric light but was apparently saved by adrenaline "forced between his lips." Ed Sullivan in his column "Broadway" reported that men had scoured Baton Rouge for adrenaline in an effort to save the life of Senator Huey Long of Louisiana after he was shot in 1935; he did receive adrenaline but died nonetheless.[12]

Airways Squeezed Tight: Adrenaline for Asthma

Asthma is a disease characterized by reversible narrowing of the small tubes that conduct air into the lungs. Severe attacks of labored breathing are triggered by marked contraction of the muscles surrounding these tubes; the narrowed tubes add considerable resistance to the flow of air in and out of the lungs. In 1900, Solomon Solis-Cohen published a seminal paper on the use of adrenal extracts in the treatment of asthma.[13] Solis-Cohen treated a twenty-two-year-old asthmatic woman with oral tablets of an adrenal extract; she quickly improved. He later reported good results in twelve other asthmatics. He did not know the exact dose to prescribe; he urged using "enough." However, it is doubtful that orally administered adrenal extracts would have raised adrenaline concentrations in the blood sufficiently for a therapeutic response; we now know that the liver has major enzymes that inactivate adrenaline before it can reach the general circulation.[14] Solis-Cohen believed that excessive dilation of blood vessels in the air tubes caused these tubes to obstruct the flow of air in some asthmatics; he used adrenaline because of its capacity to constrict blood vessels. While this reasoning proved incorrect and giving adrenal extracts orally was likely ineffective, his work stimulated considerable interest in using adrenaline in the treatment of asthma. In 1903, J. G. M. Bullowa and D. M. Kaplan in New York reported that five asthmatics experienced relief after injections of adrenaline under the skin, whereas oral treatment was ineffective.

These physicians also erroneously believed that adrenaline worked by constricting blood vessels in the walls of the air tubes.

The correct explanation for how adrenaline improves asthmatic attacks emerged about a decade later. In 1912, Edwards Park demonstrated that adrenaline dilated airways by inducing relaxation of bronchial smooth muscle. The capacity to open narrowed airways (bronchi) by inhibiting the contraction of smooth muscle in the walls of the airways is the major explanation for adrenaline's efficacy in acute asthmatic attacks; consequently, adrenaline became known as a "bronchodilator." Adrenaline injected under the skin—or intravenously in dire emergencies—became the widely accepted treatment for asthma by the 1920s.[15]

In ancient Greece, the Oracle at Delphi inhaled vapors that rose from the Earth. However, medical use of drugs suspended in air developed much later. Administering adrenaline directly into the lungs by inhalation advanced the therapy of asthma in the mid-1930s. Since adrenaline is not extensively absorbed into the blood from the lungs, this method decreased the adverse effects of adrenaline on the heart and other organs compared to injections under the skin. Inhaled adrenaline also worked very quickly. This method represented a major improvement over multiple daily injections and became quite popular with many patients.[16] (Adrenaline is still available in an over-the-counter inhaler for asthma.)

Adrenaline for Addison's Disease: An Important Failure

At the end of the nineteenth century, two general hypotheses about Addison's disease competed for supremacy: did diseased adrenals fail to secrete an active substance into the blood, or did they fail to remove poisonous substances from the blood? The Oliver and Schäfer experiments markedly stimulated ongoing efforts to use adrenal extracts to replenish needed substances. Many eminent clinicians published articles describing a few Addison's disease patients who had been treated with adrenal extracts or adrenaline. In 1896, the well-known physician William Osler reported that he had used adrenal extracts to treat six patients with Addison's disease; he believed one patient might have improved.[17] Osler had a reputation for being cautious in assessing drug efficacy and did not become a proponent of this treatment. However, many other physicians continued to use regimens containing adrenal extracts and adrenaline for patients with Addison's disease.[18]

In 1923, Bernardo Houssay provided compelling experimental evidence that the adrenal cortex, rather than the adrenal medulla, was essential for survival after removing the adrenal glands. Only in the 1930s did scientists produce sufficiently powerful adrenal cortical extracts—obtained from very large numbers of adrenal glands—to treat patients with Addison's disease effectively. The essential hormones in these extracts, cortisol and aldosterone, were isolated from the adrenal cortex over the next two decades. By the 1950s, the health of patients with Addison's disease could be restored by providing the essential hormones of the adrenal cortex.

Then and Now

When adrenaline entered the clinical arena, government had almost no role in regulating drugs, so there were essentially no obstacles to quickly marketing new medications. In modern times, innovative drugs typically emerge from fundamental biomedical research made possible by large financial investments. Many challenges must be overcome before an exceptionally promising biologically active molecule can become an approved drug. The potential drug's molecular, cellular, and physiological effects are thoroughly investigated; in addition, the chemical should prove adequately safe in animal testing. Large quantities of the chemical must be synthesized and purified for clinical trials according to stringent quality standards (good manufacturing practices). Lastly, rigorous clinical trials in humans, sometimes enrolling thousands of patients, are required to demonstrate both efficacy and safety of a new drug in the treatment of a disease. Meeting these objectives typically consumes a decade or longer; if all goes well, the candidate drug will be evaluated for approval by drug regulatory agencies around the world, such as the FDA or the European Medicines Agency. Most potential drug molecules fail to fulfill their promise at some point in this difficult journey.

Over the past hundred years, progress in drug regulation and experience gained from tragic misadventures have led to increasingly more stringent licensing requirements. Adrenaline preparations sold by Parke-Davis in the early part of the twentieth century would not satisfy current standards of good manufacturing practices. No modern regulatory agency would approve a new drug such as adrenaline without an enormous amount of formal testing in animals and demonstration of efficacy in well-defined diseases in humans.

Adrenaline in Modern Medicine

After more than a century, adrenaline remains a venerable drug in the armamentarium of physicians. However, its days as a miracle drug with overzealous admirers are long gone. While many of the purported benefits of adrenaline never panned out, adrenaline remains a drug of choice for some major clinical problems into the twenty-first century.

While no drug is capable of independently raising the dead, adrenaline is heavily used as an adjunct in cardiac resuscitation. In the United

States alone, cardiac arrest kills several hundred thousand people each year. Many of these deaths are due to ventricular fibrillation, often caused by a heart attack. In ventricular fibrillation, the heart suddenly transitions from a regular rhythm to a quivering state that cannot pump blood; blood pressure immediately drops and the victim collapses. Adrenaline is used as a component of advanced cardiopulmonary resuscitation. However, several recent observational studies have suggested that adrenaline may have adverse consequences during cardiac arrests. In 2012, in light of these concerns, an editorial in the *Journal of the American Medical Association* concluded that it would be both timely and ethical to conduct a placebo-controlled trial to find out definitively if adrenaline was either beneficial or harmful during cardiac arrests.[19]

Adrenaline remains vitally important in the treatment of life-threatening allergic reactions caused by insect stings and foods such as nuts and seafood; exposure can lead rapidly to swelling of the tongue and throat, as well as severe asthma attacks and dangerously low blood pressure. Emergency treatment with adrenaline can quickly reverse these serious manifestations of anaphylaxis.[20]

Dentists as well as ear, nose, and throat surgeons continue to use combinations of adrenaline with local anesthetics such as novocaine; as noted earlier in the chapter, adrenaline decreases bleeding and prolongs the duration of action of local anesthetics by inducing vasoconstriction.[21] In cases of massive gastrointestinal bleeding, adrenaline is used to stop bleeding from blood vessels that are visualized using special endoscopes. Adrenaline spray remains an important therapeutic option in treating children with croup—a syndrome characterized by inflammation in the back of the throat and the upper part of airways that can severely obstruct the flow of air into the lungs.

Adrenaline: Potentially Dangerous Drug

Adrenaline moved into clinical practice riding a wave of enormous optimism as a wonder drug. Nonetheless, clinicians and scientists soon drew attention to its possible adverse effects. At a 1903 meeting of the Harvard Medical Society of New York City, several physicians warned about the possibility of cardiovascular toxicity when adrenaline is infused too quickly in patients in shock. A particularly poignant case appeared in an article published in 1904 by a senior physician at the Liverpool Royal Infirmary:

> Quite recently I inadvertently induced a very severe attack of paroxysmal tachycardia in a boy who previously never had any disturbance of the kind. . . . I have been injecting adrenalin chloride solution into the serous cavities to check effusion, and lately I have been trying to find out the maximum dose for each cavity. This little fellow whom I now exhibit is suffering from very large amyloid liver and spleen, and lately, after siphoning off a large collection of ascitic fluid, I injected 10 cc of adrenalin chloride solution 1 in 1,000. This did not seem to have any immediate effect on the circulation, and unfortunately, without waiting a sufficient length of time, I injected 8 cc more of the solution. These 8 cc contracted all the arterioles of the arterial tree. . . . [T]he general arterial blood pressure rose, the heart started off at a gallop, and in a few minutes its beats numbered about 200 in the minute. . . . [T]he cardiac distress was great, and the little fellow said he was dying. . . . I cannot tell you in mm of mercury what the blood pressure was, as at the time I was otherwise occupied, but I can tell you that his little heart was taxed to its utmost capacity.[22]

One wonders what effect this experience had on the physician's future use of adrenaline.

For at least the next twenty years, opinions varied about the safety of adrenaline. For example, a 1916 editorial in the *Journal of*

Laryngology noted: "That adrenalin is regarded as a dangerous drug by many pharmacologists and anaesthetists comes with a shock of surprise to otolaryngologists, accustomed as they are to use it, locally, at all events, in a very lavish manner. For the general opinion among throat and ear surgeons undoubtedly is that adrenalin is a safe as well as a powerful remedy."

Many of adrenaline's adverse effects are direct extensions of its pharmacological actions. For example, patients treated with adrenaline may complain of palpitations. Excessive doses of adrenaline injected under the skin can so constrict blood vessels locally that the skin can be damaged by the lack of blood flow. Similarly, excessive stimulation of the cardiovascular system may lead to marked rises in blood pressure and disturbances in cardiac function including sudden death. Ironically, adrenaline can both help reverse ventricular fibrillation and cause it.

Drugs can also have adverse effects that cannot readily be anticipated. In the case of adrenaline, large injections of adrenaline in animals produce pathological, degenerative changes in arteries and in the heart that had not been anticipated. We now know that similar cardiac damage can occur in people with the adrenaline-secreting tumor pheochromocytoma, those who abuse cocaine (recall that cocaine potentiates adrenaline), and individuals under severe stress.

Adrenaline also is prominently involved in serious consequences from preventable medical errors. While the size of the problem is controversial, there is no doubt that medication errors can lead to substantial patient harm, including death.[23] For example, problems with adrenaline can arise because of labeling and dosage conventions. Vials containing adrenaline for injection are labeled by the proportion of adrenaline in the solution, typically 1:1,000, 1:10,000, and 1:100,000; these values represent concentrations of 1 milligram (mg) of adrenaline per milliliter (ml) of solution, 0.1 mg/ml, and 0.01 mg/ml, respectively. Several studies have shown that many physicians are unable to remember the correct doses of adrenaline and have considerable difficulty converting proportions such as 1:1,000 into mg/ml of drug, which is the usual way dosages for most other drugs are prescribed. In addition, vials containing different strengths of adrenaline may be placed together on shelves. Consequently, a prescribing error, using the wrong vial, or giving the drug by an unintended route of delivery (doses given intramuscularly, subcutaneously, or intravenously can affect the body in different ways) can have dire consequences.

Errors of this type with solutions of adrenaline are not a new problem; such a mistake was reported within one year of the initial marketing of adrenaline solutions by Parke-Davis. In 1902, Dr. Charles Burnett of Philadelphia reported the case of a patient in South Carolina who described her symptoms of hay fever in a letter to him. He sent her a prescription for Parke-Davis's 1:10,000 solution of adrenaline, recommending spraying three puffs into each nostril, three times daily. He wrote out the prescription in regular language, presumably trying to avoid medical hieroglyphics, and suggested that his patient seek out a "reliable druggist." However, as he described in his case report, "the druggist sent her the undiluted solution of adrenaline chloride (1 to 1000) in the original bottle, with the printed label upon it." She wrote to him again with a report of her response: "The spray . . . gives me such strange sensations. . . . My head feels full, and there is a dull, heavy pain in my forehead. . . . [T]he roaring in my ears increases . . . I have to lie down . . . [M]y hands, back and even my knees begin to tremble, and I feel as if under nervous excitement. . . . [M]y heart beats so loudly and fast that I was frightened. . . . [I]n about an hour these feelings all pass away and I feel about as usual." Fortunately, she included a copy of the drug label in her letter to Burnett, who realized that there had been a mistake and told her to stop using the solution.[24]

Opportunities for error are compounded by the need to make very rapid decisions for patients in emergency situations. Patients receiving excessive doses of adrenaline may develop severe elevations of blood pressure, cardiac disturbances, and dangerous changes in the concentration of potassium in the blood. One case report describes a young woman who inadvertently received ten times the intended dose of adrenaline for treatment of anaphylaxis associated with a tomato allergy; she suddenly developed chest pain that was later determined to be the result of adrenaline-induced takotsubo cardiomyopathy. The Pennsylvania Patient Safety Authority has taken up the call to decrease medication errors involving adrenaline (using the U.S. name "epinephrine"): "Let's Stop This 'Epi'demic!—Preventing Errors with Epinephrine." Their web site (http://patientsafetyau thority.org, consulted 29 October 2010) describes a particularly tragic death from misadventure with adrenaline. A sixteen-year-old boy was brought to an emergency department with priapism (a penile erection that persists for more than several hours). A vasoconstricting drug can be used to terminate the erection—protecting the penis from irreversible tissue damage. The treating doctor misunderstood the amount of adrenaline in a vial labeled "1:1,000"; the patient received an excessive dose of adrenaline injected into his penis. The boy's heart stopped as the adrenaline was absorbed into his circulation. This type of incident is particularly sad since it represents a well-known type of mishap with adrenaline. Almost sixty years earlier, on December 9, 1951, the *New York Times* reported that a patient in Michigan had died of an overdose of adrenaline due to a tragic error in drawing up the drug from the wrong vial during surgery for a cleft palate.

Newer Generations of Drugs Mimicking or Opposing Adrenaline

After the rush to find new clinical uses for adrenaline, some physicians gradually recognized that the drug had limitations. Despite early claims of good results after oral administration, adrenaline does not actually work when administered as a pill because it is rapidly metabolized before being absorbed into the general circulation. On the other hand, the favorable effects after it is injected or inhaled do not last long. Moreover, adrenaline causes serious adverse effects in some patients. These shortcomings suggested that orally effective, longer-acting, or more selectively acting forms of adrenaline would be clinically desirable.

Shortly after the chemical structure of adrenaline was determined, chemists began synthesizing molecules that resembled adrenaline in an effort to identify potentially more effective or safer drugs. Over time hundreds of these molecules were identified, and scientists developed a better understanding of how adrenaline activates a variety of different physiological responses. The next part of adrenaline's story covers drugs that either mimic some of adrenaline's actions or block some of its effects. Many of these improvements became possible only with major scientific discoveries in the fundamental biology of adrenaline. Chapters 6–9 describe these major biomedical advances and address how these insights provided a foundation for developing drugs that have advantages compared to adrenaline itself.

SIX

Mind the Gap:
Chemical Transmission from
Sympathetic Nerves to Organs

Everything of importance has been said before by
somebody who did not discover it.

ALFRED NORTH WHITEHEAD

When compared with injections of adrenaline, activation of sympathetic nerves has many similar actions on organs throughout the body.
The dazzling explanation for their comparable effects emerged only
after much hard work, and it had broad implications for understanding the inner workings of the nervous system and for the invention of
powerful drugs. Sorting out how the companion parasympathetic
nerves activated target organs posed an analogous challenge. The resolution of both questions involved the astonishing demonstration
that sympathetic and parasympathetic nerve endings release chemicals called neurotransmitters that move across the narrow gap from
nerve endings to cells in target organs. Since the stories leading to the
identification of these neurotransmitters are intertwined in concept,
time, place, and people, they will be discussed together.

The Concept of Chemical Neurotransmission

John Langley made great progress in understanding the physiology
of the autonomic nervous system over a fifty-year career at Cambridge
University in England. He divided the autonomic nervous system into

two components, the sympathetic nervous system and the parasympathetic nervous system. The sympathetic nerves emerge from the spinal cord largely in the chest, whereas the parasympathetic nerves originate either in the brainstem or in the low back.[1] In many cases, the effects of the sympathetic and parasympathetic nervous systems oppose each other in regulating organ function. An advantage of opposing systems is that in urgent situations, needed changes in organ function can be implemented very quickly; for example, in the case of a requirement for a rapid increase in heart rate, the accelerator of the heartbeat (sympathetic nerves) can be pushed harder as the decelerator of the rate (parasympathetic nerves) is simultaneously turned off.

Langley developed a schematic representation of the autonomic nervous system that described two key types of neurons. The first type, preganglionic neurons, is heavily regulated by signals from the brain. Preganglionic neurons send signals that are received by postganglionic neurons in structures called ganglions; the postganglionic neurons then relay the signals to the target organs. The brain uses the autonomic nervous system to regulate many functions that are outside conscious control.

In 1899, Max Lewandowski compared intravenous injections of adrenal extracts with stimulation of the sympathetic nerves going to the eye, and demonstrated that the responses in the eyes were very similar. Two years later, Langley confirmed and extended Lewandowski's results, demonstrating that the effects of adrenal extracts persisted even after the sympathetic nerves to the eye had been destroyed. This strongly suggested that the active substance in the extracts acted directly on the eyes.

In 1904, Thomas Elliott proposed a remarkable answer to a key question: how did postganglionic neurons in the sympathetic nervous system communicate with their target organs? Under Langley's supervision in Cambridge, Elliott conducted a large and insightful study comparing the effects of adrenaline with activation of sympathetic nerves.[2] He concluded that adrenaline produced physiological effects very similar to those mediated by these nerves. Elliott boldly suggested that the endings of sympathetic nerves might produce their effects in tissues by liberating adrenaline. This general concept is now called chemical neurotransmission.

At the time of Elliott's novel proposal, the electrical properties of nerves had been extensively studied, beginning more than a hundred years earlier with discoveries by Luigi Galvani and Alessandro Volta. The importance of animal electricity was increasingly well understood. Considerable experimental work had securely established that electrical currents moved down nerves, conveying signals from the brain to target organs. Physiologists widely believed that the current traveling along activated nerves jumped across the small anatomical gap from the nerves to the cells in target organs. Elliott's novel hypothesis proposed that neurons transformed an electrical signal in nerves into a chemical signal that moved from neurons to target cells. He deserves great credit for the first serious suggestion that neurotransmission involved a specific chemical.[3] Unfortunately, his daring hypothesis did not immediately stimulate much research aimed at identifying neurotransmitters. Elliott moved to London in 1906 where he completed medical training and took on other challenges.[4]

In 1906, Walter Dixon tried to extend Elliott's conjecture about neurotransmission in the sympathetic nervous system to the parasympathetic nervous system by preparing extracts of dog heart after stimulation of its parasympathetic nerves. His preliminary experiments were inconclusive, and the problem lay fallow for many years.[5]

Henry Dale and the Parasympathetic Nervous System

Henry Dale studied physiology at Cambridge and stayed on with Langley for an additional two years of research. Subsequently, he graduated in medicine from St. Bartholomew's Hospital in London.[6] He decided against a clinical career and pursued research in Ernst Starling's laboratory at University College London and in Germany with Paul Ehrlich. Even with this high-level research experience, Dale had little optimism about securing an academic job.

At the same time, Henry Wellcome was continuing his search for promising scientists to strengthen research at the Burroughs-Wellcome Pharmaceutical Company. The American pharmacists Wellcome and Silas Burroughs had founded the company in England in 1880. After Burroughs's early death, Wellcome greatly expanded the company. He offered Dale a position in the company's Physiological Research Laboratories, which would provide Dale with intellectual independence and resources for his research as well as a "marrying income." Dale boldly accepted the offer despite academics' generally low opinion of industry research, remarking, "[My] friends . . . were almost unanimous in advising me to have nothing to do with it: I should be selling my scientific birthright . . . for a mess of commercial pottage."[7]

Wellcome encouraged Dale to work on ergot extracts, complex mixtures of substances derived from a fungus that infects cereals.[8] The ergot extracts Dale tested typically raised blood pressure in animals. However, in 1913, a new batch of ergot extracts unexpectedly caused blood pressure to drop sharply. His curiosity aroused, Dale decided to identify the substance in this aberrant batch responsible for the surprising results. He and his chemist colleague Arthur Ewins purified a small amount of the unknown substance. Dale, recalling that Reid Hunt had reported eight years earlier that acetylcholine markedly lowered blood pressure, wondered if the substance under scrutiny might be acetylcholine.

In the 1860s, Adolf von Baeyer had synthesized acetylcholine from choline, but it sat on the shelf for decades.[9] In 1899, Reid Hunt, then John Abel's colleague at Johns Hopkins, began a search for novel chemicals in adrenaline-depleted adrenal extracts.[10] He found that these extracts lowered blood pressure, and he identified choline in this material. In 1906, he synthesized acetylcholine, among many other chemical analogs of choline, and found that it markedly decreased blood pressure. However, he did not demonstrate that acetylcholine existed in adrenal extracts.

Ewins compared the active substance isolated from the aberrant ergot extract with acetylcholine and found that the two molecules were identical. Dale then studied the physiological effects of acetylcholine in detail and noted their striking resemblance to the effects caused by activation of the parasympathetic nervous system. His friend Thomas Elliott had earlier stimulated Dale's interest in the concept of chemical neurotransmission. However, without evidence that acetylcholine existed in animals, Dale did not propose that acetylcholine was a possible neurotransmitter. The onset of World War I diverted Dale's thoughts to other, more pressing projects; he would return to research on acetylcholine fifteen years later with renewed enthusiasm.[11]

Otto Loewi's Remarkable Frog Experiment

In 1921, Otto Loewi made an extraordinary discovery that reinvigorated interest in chemical neurotransmission in the autonomic nervous system. The idea for his critical experiment apparently emerged under unusual circumstances:

In the night of Easter Sunday, 1921, I awoke [from a dream], turned on the light, and jotted down a few notes on a tiny slip of thin paper. Then I fell asleep again. It occurred to me at 6 o'clock in the morning that during the night I had written down something most important, but I was unable to decipher the scrawl. That Sunday was the most desperate day in my whole scientific life. During the next night, however, I awoke again, at 3 o'clock, and remembered what it was. This time I did not take any risk; I got up immediately, went to the laboratory, made the experiment on the frog's heart.[12]

Loewi placed a frog heart in a beaker filled with a salt solution and stimulated the heart's parasympathetic nerve (vagus nerve); the heart slowed. He then transferred the solution surrounding the heart to a second heart. That heart also slowed, mimicking the effect of parasympathetic nerve stimulation. Similarly, when he stimulated the accelerator (sympathetic) component of the nerve to increase heart rate, the solution from that heart increased the rate of a fresh heart. These results suggested that parasympathetic and sympathetic nerves in the heart released substances into the salt solution that decreased or increased heart rate, respectively. He tentatively named them *Vagusstoff* (vagus substance) and *Acceleransstoff* (accelerator substance).

Elliott had made the brilliant suggestion of chemical transmission in the sympathetic nervous system but did not pursue the problem. Henry Dale had brushed up against a similar concept with his acetylcholine experiments. Then Loewi confirmed the general concept in an elegant experiment.[13] Henry Dale later described the impact of Loewi's famous experiment: "Transmission by chemical mediators was like a lady with whom the neurophysiologist was willing to live and to consort in private, but with whom he was reluctant to be seen in public. And then, all that was changed when Loewi made his straightforward experiments, stood this egg of Columbus upright on its flattened end, and gave us experimental facts in place of half-discredited speculations."[14] Nonetheless, Loewi's findings were strongly criticized for more than a decade.

Some skeptical scientists could not reproduce the results, doubted the relevance to humans of data from frogs, or disagreed with the interpretation of the experiments. Bruno Minz wrote about the Swiss physiologist Leon Asher's vigorous criticism of Loewi: "Two respectable scientists attack one another in terms which could only have escaped the mouths of Homer's heroes. Asher even suggested that Loewi's thinking was a little 'autistic' by definition of Blueler [i.e., schizophrenic]." In an effort to convince other scientists of the validity of his results, Loewi publicly demonstrated his frog experiments at the International Physiology Congress in 1926 in Stockholm. At the beginning of the 1930s, Minz divided interested physiologists into five groups: people who could repeat Loewi's experiments and believed the interpretation; others who could repeat them but did not accept the interpretation; those who failed to reproduce the results; those who never tried the experiments but believed them; and those who never tried them and were unconvinced.

The Unmasking of Vagusstoff

Loewi found that atropine antagonized the slowing of heart rate by both acetylcholine and *Vagusstoff*. He also found that stimulating the heart's parasympathetic nerve increased the concentration of choline around the heart. While choline itself did not slow the heart rate, Loewi wondered if it might be a breakdown product of acetylcholine. He also discovered that physostigmine prolonged the action of both *Vagusstoff* and acetylcholine. We now know that physostigmine inhibits acetylcholinesterase, the enzyme that rapidly breaks down acetylcholine. Consequently, physostigmine allowed both *Vagusstoff* and acetylcholine to accumulate.[15] Loewi's experimental results led him to believe that *Vagusstoff* was acetylcholine. His findings were highly suggestive but not definitive, since he lacked a method to accurately measure acetylcholine.

In 1929, Henry Dale and Harold Dudley demonstrated for the first time that acetylcholine existed in animal tissues.[16] With this new foundation, Dale speculated explicitly that acetylcholine might be physiologically important. However, he could not conclude that acetylcholine was *Vagusstoff* in the absence of compelling evidence that parasympathetic nerves released acetylcholine. Three years later, Dale learned about a new assay for acetylcholine and recognized that this method might be sensitive enough to detect acetylcholine released from nerve endings.

The idea for the assay stemmed from work of Hermann Fühner, who in 1918 demonstrated that muscles in leeches contracted in response to acetylcholine; adding physostigmine made the muscles about a million times more sensitive to acetylcholine. Bruno Minz and Wilhelm Feldberg used Fühner's findings to design a novel assay for acetylcholine. They exposed leech muscle to biological fluids containing unknown amounts of acetylcholine and measured contractions. They then compared these responses to contractions obtained with known concentrations of chemically synthesized acetylcholine.

At about the same time, both Minz and Feldberg lost their positions because of Nazi laws restricting employment of Jews in Germany, and Dale recruited Feldberg to his laboratory in England.[17] Before leaving for England, Feldberg, with Otto Krayer, used the leech assay to demonstrate that stimulation of the heart's parasympathetic nerves increased the amount of acetylcholine in blood draining from the heart.[18]

Feldberg quickly implemented the leech assay in Dale's laboratory.[19] Feldberg, John Gaddum, and Dale used the assay to demonstrate that postganglionic parasympathetic nerve endings released acetylcholine. In addition, they demonstrated that preganglionic neurons in the autonomic nervous system also released acetylcholine at the connections with postganglionic neurons in ganglions. Dale, Feldberg, and Marthe Vogt demonstrated that motor nerves also released acetylcholine in skeletal muscle.[20] These results strongly suggested that acetylcholine was the neurotransmitter at these three locations.[21]

The Unmasking of Acceleransstoff

Identifying *Acceleransstoff*, the sympathetic nervous system neurotransmitter, involved a tough, tortuous journey that challenged both imagination and assay sensitivity. In addition to Loewi's seminal frog experiments, support for the idea of chemical transmission in the sympathetic nervous system came from Walter Cannon at Harvard Medical School, who with collaborators such as Zenon Bacq demonstrated that sympathetic nerves released a substance into the blood that mimicked activation of the sympathetic nervous system. Cannon called this substance *sympathin*. In 1930, Cannon and Arturo Rosenblueth found that adrenaline incompletely mimicked the effects of sympathin, and they concluded that the sympathetic nervous system's neurotransmitter was similar to but distinct from adrenaline.[22]

What was sympathin? Possibly related to this question was the unsolved problem of how adrenaline and the sympathetic nervous system could, depending on the tissue, induce either the contraction or relaxation of smooth muscle, a finding extensively documented in Thomas Elliott's work in 1904. Cannon and Rosenblueth proposed a complex, multifaceted hypothesis addressing both problems. The proposal stated that adrenaline became active only after combining with one of two factors in tissues; they named these factors E (excitatory) and I (inhibitory). Adrenaline-E caused excitatory responses such as smooth muscle contraction, whereas adrenaline-I had the opposite effect. Sympathin-E and sympathin-I had analogous actions. Promulgated by the greatest physiologist in the United States, this complex proposal attracted considerable attention. However, many scientists found the hypothesis baroque and lacking in biochemically based supporting evidence. Despite continued criticisms, Rosenblueth held tenaciously to this conjecture and Cannon loyally did not readily give ground. The resulting controversy possibly singed Cannon's otherwise impeccable scientific reputation. In the long run, their hypothesis was refuted. Sympathin-E and sympathin-I did not exist; rather, the neurotransmitter in the sympathetic nervous system interacted with receptors on target cells that could have excitatory or inhibitor effects depending on the tissue (see Chapter 8 for the story of the discovery of these receptors).

Additional clues about sympathetic neurotransmission came from other directions. In 1904, Friedrich Stolz synthesized noradrenaline, which initially attracted little interest from physiologists. However, in 1910, George Barger and Dale studied the effects of noradrenaline along with many other compounds structurally related to adrenaline. They coined the term *sympathomimetic* for chemicals that reproduced some or all of the effects of the sympathetic nervous system. They learned that noradrenaline mimicked the effects of the sympathetic nervous system more faithfully than did adrenaline. Years later, Dale retrospectively chastised himself for not interpreting his results as a subtle hint that noradrenaline was a better candidate than adrenaline as a possible neurotransmitter. In 1934, Zenon Bacq suggested that the sympathetic nervous system transmitter might be noradrenaline (among other possibilities), but he did not produce any compelling experimental evidence in support of this conjecture. In 1935, in a major lecture to the Royal Society in London, Loewi expressed conviction that the parasympathetic nerves released acetylcholine; however, he only went as far as saying that the sympathetic neurotransmitter in the frog heart was "adrenaline-like."[23]

Other scientists had results compatible with the hypothesis that noradrenaline was the sympathetic neurotransmitter, but none had a sufficiently sensitive assay to make definitive measurements.[24] Peter Holtz in Germany and Hermann (Hugh) Blaschko in England independently found suggestive evidence that noradrenaline existed in the body.[25] In 1944, Holtz submitted a paper (published only in 1947, as it was delayed by the war) demonstrating the presence of noradrenaline in the adrenal medulla and urine. Using ingenious bioassays and a novel chemical assay, Ulf von Euler shortly afterward provided compelling measurements establishing that noradrenaline was in fact the major neurotransmitter in sympathetic neurons.[26]

Ulf von Euler's Journey to Sympathetic Neurotransmission

Ulf von Euler came from a scientifically distinguished family; his father, Hans von Euler-Chelpin, shared the 1929 Nobel Prize in Chemistry for work on fermentation of sugars, and his mother, Astrid Cleve, was the first Swedish women to receive a PhD in a science (botany).[27] As a student, von Euler attended the 1926 physiology congress at which Loewi demonstrated his frog heart experiments. After research training in Sweden, von Euler contributed to the work on acetylcholine in Henry Dale's laboratory; he also collaborated with Gaddum in the discovery of substance P, the first member of a now large family of important peptides. This discovery made a profound impression on von Euler: "These early experiments not only whetted the appetite for finding more active substances in biological material but also provided the necessary 'know-how' for making such attempts successful." He next worked with Corneille Heymans (Nobel Prize in Medicine, 1938). (Nobel Prize in Physiology or Medicine is abbreviated to Nobel Prize in Medicine throughout the book.) Just before the onset of World War II, Von Euler again worked in Dale's laboratory.

In 1934, von Euler serendipitously discovered a new world of chemical mediators while investigating an adrenaline-like substance in extracts of prostate glands. He extended these experiments from the prostate to the nearby seminal vesicles—glands whose fluids contribute to the formation of semen along with sperm and secretions from the prostate. Von Euler injected human seminal fluid (of unknown provenance) into rabbits and remarked humorously on his excitement at the results: "The blood pressure . . . dropped to very low values . . . while my own probably went up some." But his euphoria vanished when he learned that investigators in New York and London had recently published on biological activity in seminal fluid.[28] Von Euler named the unknown active substance *prostaglandin*, presumably because he found activity in the prostate gland. Von Euler began purifying prostaglandin from sheep seminal vesicle material and made some progress, but with the onset of World War II he moved on to other research problems.[29]

Von Euler was keenly interested in identifying the neurotransmitter released by postganglionic sympathetic neurons; he wanted to solve the problem by biochemically analyzing nerve and tissue extracts. He later described his focus on the sympathetic neurotransmitter, displacing other very promising research involving substance P and prostaglandins: "It must also be remembered that at this time research funds were very limited and teaching took a good deal of the available time. It was necessary to select one field and not split the resources on several topics." In 1945, he wrote Dale that he was working hard in his laboratory after getting relief from teaching responsibilities.

Von Euler purified a substance from cattle spleen that had pharmacological properties similar to those of noradrenaline; he found this substance in many other tissues, including sympathetic nerves. Evidence from many different experimental approaches supported the role of noradrenaline as the adrenergic neurotransmitter. Over the course of a decade, scientists progressively recognized noradrenaline as the long-sought neurotransmitter in the sympathetic nervous system. Henry Dale wrote in his introduction to von Euler's 1955 book *Noradrenaline*, "There can seldom have been a case in which a substance [noradrenaline] . . . has attained so rapid a celebrity . . . in so short a period, to so large a volume of research by so many experienced investigators." Dale also reflected on his famous paper with Barger, published in 1910, demonstrating that noradrenaline mimicked the effects of sympathetic nervous system activity better than adrenaline:

> I find it curious, even a little humiliating, to reflect that the main significance which I found in this observation at that time, was that it seemed to make it difficult to accept the brilliant hypothesis which had been put forward some 5 years earlier by my friend, TR Elliott, attributing to adrenaline the function of a chemical transmitter of the effects of sympathetic-nerve impulses. It seems so obvious in the light of all that has happened since Otto Loewi produced the first direct evidence for such a function in 1921 that we ought to have recognized immediately the possibility that Elliott's hypothesis might still be right in principle, and that the next step should have been to test, even in 1910, the possibility that . . . [noradrenaline] might even, at many nerve-endings, be the predominant component of the transmitter.

Several important discoveries flowed from improved methods to precisely measure noradrenaline and adrenaline. In Oxford, Edith

Bülbring demonstrated that noradrenaline was the precursor to adrenaline in the adrenal medulla.[30] In 1948, William (Owen) James in the Department of Botany at Oxford used the new technique of paper chromatography to definitively separate noradrenaline from adrenaline. Several groups made the surprising discovery that supposedly pure medicinal-grade adrenaline, derived from animal adrenals, was actually 30 percent "contaminated" by noradrenaline.[31]

Sequel to the Theory of Chemical Neurotransmission

In 1936, Henry Dale and Otto Loewi shared the Nobel Prize in Medicine for discovering that acetylcholine is the neurotransmitter in the parasympathetic nervous system. John Eccles, a strong proponent of electrical theories of neurotransmission, remained a tenacious critic of acetylcholine as neurotransmitter. His intellectually rigorous objections stimulated advocates of acetylcholine to strengthen their case. Complete acceptance of chemical transmission throughout the nervous system occurred only in the 1950s, when more advanced microscopic and electrophysiological techniques powerfully excluded alternative hypotheses. In 1970, von Euler received a share of the Nobel Prize for Medicine for demonstrating that noradrenaline is the major neurotransmitter in postganglionic sympathetic nerve endings. With the paradigm of chemical neurotransmission firmly in place as the mode of communication used by neurons to signal innervated cells, a host of additional neurotransmitters were identified in other neural pathways, including small molecules such as glutamate and glycine as well as many short peptides. Indeed, postganglionic sympathetic nerve endings also release adenosine-5'-triphosphate (ATP) and peptides, which may modulate some of the effects of noradrenaline by activating their own specific receptors.

The Origin and Fate of Adrenaline and Noradrenaline in the Body

The discoveries of adrenaline and later noradrenaline raised questions about how they were synthesized. What was the starting material, and what were the chemical steps leading to the production of each substance? The very short duration of action of adrenaline and noradrenaline suggested that these substances disappeared quickly; what was their fate in the body?

Biosynthesis of Adrenaline

The combined efforts of many investigators over several decades elucidated how adrenaline is synthesized in the body. The starting material is tyrosine, a naturally occurring amino acid found in proteins consumed as food; the pathway has four key steps.

In sympathetic nerve endings, synthesis stops at the third step, resulting in noradrenaline; in the adrenal medulla, an additional step produces adrenaline. The discoveries of the various enzymes in the pathway constituted major biochemical milestones. As early as 1906, speculation flourished that adrenaline arose from chemical modifications to tyrosine because of their close structural resemblance. Since four chemical modifications are needed, there are at least twenty-four ($4! = 4 \times 3 \times 2 \times 1 = 24$) hypothetical sequences between tyrosine and adrenaline. In 1939, Blaschko suggested the correct synthetic pathway, although it took decades to confirm this conjecture as well as identify the enzymes that catalyzed each step.

Tyrosine hydroxylase, the first enzyme in the pathway, was isolated last. In 1964, Sidney Udenfriend and Albert Sjoerdsma demonstrated that metyrosine inhibited the enzyme, blocking the conversion of tyrosine to DOPA; metyrosine is still used to block the synthesis of noradrenaline and adrenaline in some patients with

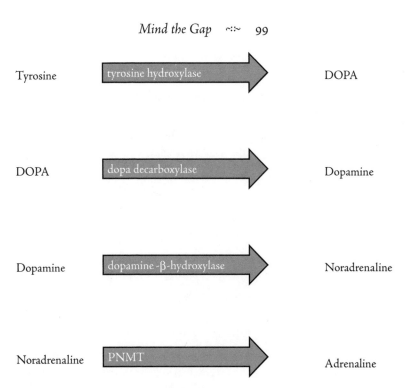

Figure 6-1. The synthesis of adrenaline. Starting with the amino acid tyrosine, contained in proteins found in many foods, adrenaline is synthesized in four sequential steps, dependent on four separate enzymes to catalyze the reactions. Dopamine is an important neurotransmitter in its own right, especially in the brain. Noradrenaline is the major transmitter in postganglionic sympathetic neurons. The synthetic pathway continues on to adrenaline primarily in the adrenal medulla. DOPA is 3,4-dihydroxyphenylalanine; PNMT is phenylethanolamine N-methyl transferase.

pheochromocytomas. Just before World War II, Holtz discovered the enzyme DOPA decarboxylase (now called aromatic L-amino acid decarboxylase), which converts DOPA to dopamine.[32] Dopamine, noradrenaline, and adrenaline are catecholamines, substances that contain both a catechol moiety (a functional group that is part of a molecule) and an amino group.[33] Seymour Kaufman made many important contributions to the purification of dopamine-β-hydroxylase, the enzyme that converts dopamine to noradrenaline.[34] In 1961, Julius Axelrod purified phenylethanolamine N-methyl transferase, the enzyme that converts noradrenaline to adrenaline by adding a methyl group to noradrenaline. This enzyme is mainly found in the adrenal medulla. The adrenal medulla is bathed in blood that has passed through the adrenal cortex. Consequently, chromaffin cells in the medulla are exposed to undiluted concentrations of cortisol, an essential hormone made in the adrenal cortex. This intimate anatomical connection provides an exceedingly high concentration of cortisol to the medulla, an essential requirement for maintaining expression of phenylethanolamine N-methyl transferase and ongoing synthesis of adrenaline.

Turning Off Adrenaline and Noradrenaline

The sympathetic nervous system is designed to not only turn on rapidly but also turn off quickly. The turn-off mechanisms for adrenaline and noradrenaline are multifaceted. Two enzymes play major roles in converting catecholamines into inactive metabolites. In addition, specialized pumps rapidly recycle noradrenaline back into sympathetic neurons, a move that plays a major role in ending noradrenaline's actions.

Metabolism of Adrenaline

The enzyme monoamine oxidase (MAO) transforms adrenaline and noradrenaline to inactive metabolites by deamination (removal of the nitrogen-containing amino group). The discovery of this enzyme involved important independent contributions from Mary Hare-Bernheim, Blaschko, and Juda Quastel.[35] MAO plays a vital role in protecting the body from absorbing treacherous chemicals from food that would otherwise disrupt the sympathetic nervous system. MAO likely inactivated most if not all of the adrenaline in adrenal extracts that George Oliver administered orally in his human experiments.

In 1957, Julius Axelrod discovered an entirely distinct enzyme that also inactivates adrenaline and noradrenaline. Called catechol-O-methyltransferase (COMT), this enzyme adds a methyl group to the 3-hydroxyl group on adrenaline and noradrenaline; the resulting metabolites are biologically inactive.[36] Measurements of these metabolites in blood are very useful in diagnosing pheochromocytoma.

Uptake of Adrenaline and Noradrenaline

With two enzymes capable of inactivating adrenaline and noradrenaline, the explanation for the turn-off mechanism seemed at hand. However, Axelrod found that even after inhibiting the activity of both COMT and MAO, the effects of noradrenaline nonetheless rapidly terminated. He discovered an additional turn-off mechanism that powerfully recycles about 90 percent of noradrenaline released by sympathetic neurons. Noradrenaline is quickly pumped back into the nerve endings, remaining as good as new.[37] Noradrenaline that escapes recycling is ultimately inactivated by enzymes.

Extensions of these experiments provided the explanation for Loewi's discovery that cocaine magnified the effects of adrenaline. Cocaine blocks the pump that recycles noradrenaline back into neurons; in the presence of cocaine, noradrenaline reaches unusually high concentrations near target tissues, leading to marked arterial constriction and rises in blood pressure. In the brain, the inhibition of dopamine reuptake into neurons plays a major role in inducing the well-known pleasurable effects of cocaine use.

A specific transport pump carries noradrenaline back into nerve endings. Three decades after Axelrod's work, scientists isolated the gene that encodes the noradrenaline transporter protein; this discovery was followed by the identification of the genes for transporters primarily responsible for pumping the neurotransmitter serotonin back into neurons. The antidepressant drug fluoxetine (trade name Prozac), the first major selective serotonin reuptake inhibitor (SSRI), works by inhibiting the reuptake of serotonin into neurons. Other antidepressant drugs inhibit both serotonin and noradrenaline reuptake into neurons; venlafaxine (trade name Effexor) is an example of a serotonin-noradrenaline reuptake inhibitor. For his work on the reuptake and metabolism of catecholamines, Axelrod received a share of the 1970 Nobel Prize in Medicine.

Julius Axelrod's life story is inspirational, especially for young scientists who are neither child prodigies nor silver-spoon graduates of exclusive universities. Brought up in a poor family on the Lower East Side of New York, Axelrod graduated from the tuition-free City College of New York. He applied unsuccessfully to multiple medical schools. In 1933, he became a laboratory technician at New York University. A few years later, he moved to the Laboratory of Industrial Hygiene, measuring vitamins in foods; he worked on a master's degree in chemistry at night.

For a special project investigating adverse effects of analgesic drugs, Axelrod fortuitously had the opportunity to collaborate with Bernard "Steve" Brodie, an accomplished pharmacologist.[38] Interactions with Brodie transformed Axelrod's scientific outlook; he learned that doing important research was very satisfying. Brodie and Axelrod discovered a novel drug metabolite with analgesic activity; this substance, acetaminophen, later became a major over-the-counter medication for pain (trade name Tylenol). Brodie and Axelrod moved to the National Institutes of Health (NIH), where Axelrod became interested in sympathomimetic drug metabolism, beginning with ephedrine (the active ingredient in ma huang) and amphetamine. This work led to the discovery of a novel class of enzymes in the liver that are critical in the metabolism of many drugs. In addition, he studied the metabolism of caffeine and LSD and did research on cocaine and marijuana. When asked if he was ever tempted to try any of these drugs, he replied, "I think you'd be crazy to do it . . . I get my kicks from doing research."[39]

SEVEN

How Adrenaline Stimulates Cells

There are also unknown unknowns—the ones we don't
know we don't know.

DONALD RUMSFELD

In the first several decades of the twentieth century, physiologists identified the most prominent effects of adrenaline on major organs such as the heart and lungs. Elucidating the mechanisms underlying these actions presented highly attractive scientific problems that proved very difficult to solve. The explanations required discovering and unraveling intricate, submicroscopic actions that adrenaline has on specific molecules in cells that make up organs. This broad field of research is now known as signal transduction biology, the investigation of how the arrival of a signal such as adrenaline at cells is transduced biochemically into responses within cells. There was no early indication that figuring out how adrenaline affected cellular function would reveal how many other apparently unrelated drugs and hormones had their effects.

A curious observation that initially had no special importance provided an important direction for research that later ignited major insights. In 1901, Ferdinand Blum in Germany reported that injections of adrenaline caused glucose to appear in the urine.[1] Georg Zuelzer soon followed by demonstrating that glucose spilled into the urine because adrenaline increased glucose concentrations in the blood. Experiments aimed at gaining a fuller understanding of adrenaline's effects on glucose metabolism proved exceedingly fruitful in identifying fundamental biochemical actions that adrenaline has in cells throughout the body.

Adrenaline Regulates Enzyme Activity: Liberation of Glucose from Glycogen

Carl Cori was born in Prague, then part of the Austro-Hungarian Empire.[2] As a student, he did research on adrenaline and heart rate. Shortly after graduation in 1920, he married a classmate, Gerty Radnitz, with whom he had done research as a student.

Finding work in post–World War I Vienna proved very difficult, so in 1921 Carl accepted a temporary research position with Otto Loewi in the Pharmacology Institute in Graz and decided to focus his research on the fate of glucose in the body. Desperately seeking posts at a time when conditions were deteriorating in Europe, the Coris even applied for medical positions in colonial Indonesia. In the meantime, an offer arrived from the State Institute for the Study of Malignant Disease in Buffalo, New York (now Roswell Park Cancer Institute), which they accepted. In 1922, Carl took on management of the clinical blood test laboratory and Gerty did microscopy of clinical tissue samples; they conducted research in their spare time.[3]

The major thrust of the Coris' research program involved studying the regulation of glucose metabolism. Powerful biological controls ensure that glucose concentrations do not rise or fall excessively with eating or fasting, respectively. Failure of these regulatory mechanisms can lead to abnormal glucose concentrations; for example, the hallmark of diabetes is high glucose concentrations in the blood.

In the 1850s, Claude Bernard demonstrated that glucose enters the blood from two main sources: from the intestines after the digestion of carbohydrates in food, or, during fasting, from the liver, which releases glucose into the blood from glycogen stores. As the Coris launched their research, scientists knew that insulin and adrenaline had opposing effects on glucose: insulin lowered the glucose concentration in blood, while adrenaline increased its concentration. The Coris focused on the role of glycogen in the regulation of glucose metabolism.[4] They demonstrated that adrenaline stimulates the breakdown of glycogen, chopping off molecules of glucose. In skeletal muscles, this glucose is used to produce chemical energy for the work of muscle contraction. Additionally, the Coris demonstrated that in the liver, adrenaline both increases the liberation of glucose from glycogen and increases the secretion of glucose into the blood. Consequently, adrenaline enhances the delivery of glucose to other organs, providing them with a source of chemical energy. This is an important component of the fight-or-flight response.

Between 1922 and 1930, the Coris made considerable progress in understanding the physiological effects of adrenaline on glycogen in liver and muscle in intact experimental animals.[5] At this point, the Coris boldly decided to focus their efforts on unraveling the mechanisms responsible for the physiological effects they had discovered. Many investigators cautiously and with little risk continue indefinitely on well-beaten paths, almost certain to find new information but with little likelihood of making major discoveries. The Coris' decision to transition from highly successful physiological experiments to new and very technically challenging biochemical investigations represented superb judgment and dedication to pursuing important problems wherever they led.

The Coris conducted experiments in progressively more simplified systems—moving from intact animals to experiments using isolated pieces of muscle and liver, and then studying enzymes arduously isolated from these tissues. In general, biochemical investigations are more powerful when they are conducted in simplified model systems that allow for much greater experimental control. The Coris fully exploited their focus on isolated tissues and enzymes after moving to the dynamic research environment at Washington University in St. Louis in 1931.[6] Carl became chair of pharmacology, and Gerty initially had a research position with a small salary, only later becoming a professor herself.

In St. Louis, the Coris focused on a fundamental problem: how does adrenaline foster the liberation of glucose from glycogen? An enzyme called glycogen phosphorylase cuts glucose loose from glycogen's long chains. The Coris discovered that adrenaline activated this enzyme, increasing the cutting rate. Exactly how adrenaline activated glycogen phosphorylase was mysterious. With considerable scientific courage, the Coris decided to purify this enzyme from crude tissue extracts containing hundreds of other proteins in the hope of gaining insight into its activation. They found two forms of the enzyme: an active form, phosphorylase a, that liberated glucose from glycogen, and an inactive form, phosphorylase b. Carl Cori and his graduate student Earl Sutherland demonstrated that adrenaline and the hormone glucagon activated glycogen phosphorylase by enhancing the conversion of inactive phosphorylase b to active phosphorylase a. These experiments represented an enormous scientific advance: somehow adrenaline activated an enzyme inside cells.

The solution of a good scientific problem often poses important new questions, much as climbing to the top of a mountain may bring new, challenging vistas into view. In this case a key challenge was to determine the chemical difference between the active and inactive forms of glycogen phosphorylase and determine how the change came about. Solutions to these problems would originate with graduates of the Cori laboratory.

In 1947, Carl and Gerty Cori shared half the Nobel Prize in Medicine for their discoveries involving enzymes and glycogen (Bernardo

Houssay received the other half for work on a pituitary hormone that influenced the metabolism of glucose); Gerty Cori became the first American woman awarded a Nobel Prize.[7]

Gerty and Carl Cori's accomplishments remain inspirational. As broadly educated physicians with strong backgrounds in physiology, they had the perspective to connect important metabolic processes in the body to fundamental biological questions.[8] They had the drive and courage to pursue important questions with novel and challenging experiments to reach fundamental explanations, working with the highest standards of intellectual rigor and commitment. And they served as formidable teachers of the next generation of biomedical scientists.

Molecular Basis for the Activation
of Glycogen Phosphorylase

In the mid-1950s, the elucidation of the molecular differences between active and inactive glycogen phosphorylase emerged independently from the laboratories of two former members of the Cori laboratory—Earl Sutherland in Cleveland with Theodore Rall, and Edwin Krebs in Seattle with Edmund Fischer. The two forms of phosphorylase differ only by a single small chemical group. Glycogen phosphorylase a, the active enzyme, contains a phosphate group; removal of this phosphate group produces glycogen phosphorylase b. Krebs and Fischer purified a novel enzyme from muscle that phosphorylated inactive glycogen phosphorylase b, converting it to the active form glycogen phosphorylase a. This remarkable enzyme became known as phosphorylase kinase. Sutherland and Rall identified an enzyme that inactivated glycogen phosphorylase a by removing its phosphate group; enzymes that remove phosphate groups are known as phosphatases. We now know that one specific amino acid in the chain of more than eight hundred amino acids in muscle glycogen phosphorylase gets phosphorylated. The discovery that adding a small phosphate group to a huge protein molecule could radically change its enzymatic activity was very surprising. Subsequent research demonstrated that phosphorylation/dephosphorylation reactions underlie many of the most fundamental biological processes, ranging from the action of numerous drugs and hormones to cell division in both normal and malignant cells.

With the knowledge that the phosphorylation of glycogen phosphorylase determined enzyme activity, a critical question emerged: how does adrenaline increase the phosphorylation of glycogen phosphorylase? This problem had an especially elegant solution that led to a fundamentally new paradigm about how hormones activate cellular responses. The full explanation required several major breakthroughs that have had broad implications; these discoveries involved unearthing a substance known as cAMP and identifying a protein kinase that is activated by cAMP.

Discovery of cAMP

In the late 1950s, Sutherland and Rall discovered that adrenaline-mediated activation of glycogen phosphorylase depended upon the release of a small chemical from the cell surface membranes that enclose the contents of every cell. They succeeded in purifying a tiny amount of this material and found that the substance contained the chemical building blocks adenine, ribose, and phosphate. Serendipity coupled with free exchange of scientific information helped speed the way to solving its exact chemical structure. Sutherland wrote Leon Heppel, a noted expert in nucleotide chemistry, asking for a reagent to help continue this work. Heppel had earlier received a letter from David Lipkin at Washington University about collaborative research with Roy Markham that had produced a puzzling compound

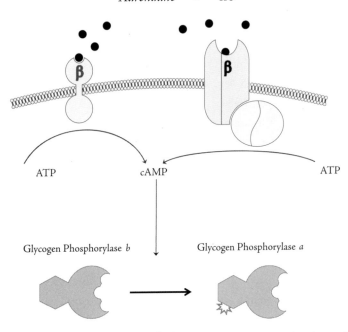

Figure 7-1. Adrenaline (●) binds to β adrenergic receptors, increasing the production of cAMP by the enzyme adenylyl cyclase. Whether the receptor and adenylyl cyclase were one protein molecule (left) or two separate molecules (right) was unknown at the time of the discovery of cAMP. cAMP transforms inactive glycogen phosphorylase b to active glycogen phosphorylase a in liver and muscle cells.

by heating ATP in a solution of barium hydroxide. Heppel happened to reread these two letters on the same day—one account has it that he found them in the same pile of mail, another that he retrieved them from a folder holding his tax return receipts—and based on tentative ideas contained in each letter, Heppel decided that the two groups might be working with the same compound. Sutherland and the investigators in St. Louis exchanged compounds; the availability of large amounts of chemically synthesized material in Lipkin's laboratory facilitated the investigations. In fact, the two groups were working with identical substances. The chemical structure was adenosine 3', 5'-cyclic monophosphate (initially abbreviated to cyclic AMP and now generally called cAMP).

Sutherland and Rall demonstrated that cAMP activated glycogen phosphorylase in the absence of adrenaline. They also discovered that adrenaline stimulated an enzyme—now called adenylyl cyclase—that converts ATP to cAMP; this enzyme was found in the membranes that enclose cells. These were monumental discoveries that shed new light on how adrenaline acted on cells. Half a century before, Ernest Starling had introduced the term *hormone* to describe specialized substances serving as chemical messengers. Adrenaline is the initial messenger from the adrenal medulla that informs the liver to release glucose; Sutherland christened cAMP a "second messenger" produced in response to the "first messenger," adrenaline.[9] He proposed that cAMP somehow triggered biochemical responses inside cells. (See Figure 7-1.)

Sutherland and others detected cAMP in cells throughout the body. Revolutionary experiments demonstrated that many hormones and drugs unrelated to adrenaline increased cAMP concentrations in a wide variety of cells, and it eventually became known that cAMP has many physiological effects in addition to activation of glycogen phosphorylase. Consequently, cAMP is a remarkably versatile second messenger, mediating many of the effects of numerous drugs and hormones. In addition, the generality of Sutherland's concept of second messengers increased with the later recognition that calcium and specific lipid molecules can also function as second messengers in signaling mediated by adrenaline and other drugs.

Sutherland began research as a medical student in Carl Cori's laboratory working on glucose metabolism. After graduation, Sutherland served in World War II as a battalion surgeon in Patton's army. In 1945, he returned to his research with Cori. In 1953, Sutherland moved to become the chair of pharmacology at Western Reserve University.

Sutherland's conviction that he could develop a simplified, cell-free system to study hormone activation of glycogen phosphorylase led to the discovery of cAMP. He revolutionized the understanding of hormone action by discovering cAMP, for which he received the Nobel Prize in Medicine in 1971. His colleagues remembered him as a kind, independent man with superb scientific intuition and as a master of experimental design.

How Does cAMP Stimulate Phosphorylation of Glycogen Phosphorylase?

The discovery of the second messenger cAMP raised the pressing question of how cAMP triggered activation of glycogen phosphorylase. Starting in the late 1950s, Krebs and Fischer collaborated to solve this fundamental problem. They found that cAMP activates a novel protein kinase called cAMP-dependent protein kinase (shortened now to protein kinase A or PKA). PKA in turn activates phosphorylase kinase by phosphorylation. Activated phosphorylase kinase then phosphorylates glycogen phosphorylase. This mouthful of enzyme names should not disguise the elegant pathway that muscle and liver cells use to convert the arrival of adrenaline into the activation of glycogen phosphorylase and the liberation of glucose from glycogen stores: adrenaline increases accumulation of cAMP; cAMP activates PKA, which in turn activates phosphorylase kinase, which in turn activates glycogen phosphorylase. The combined work of Krebs and Fischer, along with that of Sutherland, unveiled the complex cascade of events that intervened between adrenaline's interaction with its receptors and the liberation of glucose from glycogen.

Having multiple steps in a signaling pathway provides several opportunities to amplify biological signals. Tiny increases in adrenaline in the blood trigger the production of many cAMP molecules inside target cells; each cAMP molecule contributes to activating many

PKA molecules, which in turn phosphorylate many target protein molecules. Consequently, a small change in adrenaline concentration can trigger considerable action inside cells. Using phosphorylation to control enzyme activity advantageously facilitates rapid on-and-off mechanisms; this is a remarkably efficient recycling mechanism compared to the synthesis and degradation of large enzymes.

PKA has two different subunits. Its regulatory subunits recognize cAMP, and its catalytic subunits phosphorylate proteins. The complete PKA enzyme is composed of two regulatory subunits and two catalytic subunits. Binding of cAMP to the regulatory subunits liberates the catalytic subunits; the free catalytic subunits phosphorylate target proteins. The catalytic subunit of PKA is associated with several prominent milestones in protein chemistry: PKA was the first protein kinase to have its amino acid sequence determined, and the first protein kinase to have its three-dimensional structure resolved by X-ray crystallography. We now know that there are four different genes for the regulatory subunits and two genes for the catalytic subunits.[10] While the vast majority of effects of cAMP are mediated by PKA, cAMP has a few additional effects in cells that do not involve activation of PKA.

Phosphorylation pathways were initially considered special to glucose metabolism. Subsequent discoveries demonstrated the very broad, general significance of phosphorylation in cellular regulation across all of biology. For example, adrenaline stimulates phosphorylation of an enzyme (lipase) in fat cells that liberates fatty acids into the blood for use as fuel in other tissues. Adrenaline increases the strength of contraction in the heart by PKA-mediated phosphorylations. Hundreds of different protein kinases have now been identified. These kinases are activated by second messengers including cAMP, calcium, lipid molecules, and other kinases as part of cascades of kinases. Phosphorylation does more than control enzyme activity: it is involved in regulation of gene expression, movement of proteins within cells, and targeting of proteins for destruction.[11] Many malignancies are caused by excess activation of specific protein kinases; drugs that inhibit these kinases have important roles in the treatment of cancer. The burgeoning field of protein phosphorylation was launched by Krebs and Fischer's discoveries of glycogen phosphorylase kinase and PKA. For their discoveries concerning reversible protein phosphorylation as a key biological regulatory mechanism, Krebs and Fischer shared the 1992 Nobel Prize in Medicine, more than three decades after some of their major discoveries.

Krebs studied medicine at Washington University and after graduation started a residency in internal medicine, which was interrupted to serve as a physician in the Navy during World War II. Returning to Washington University, Krebs intended to complete clinical training, but a long waiting list for residency positions after the war led him to the Cori laboratory. He worked on glycogen phosphorylase in muscle and became hooked on research, deciding to forgo further clinical training. In 1948, he joined the faculty at the University of Washington in Seattle.[12]

Fischer was born in Shanghai, China, where his journalist father had moved from Vienna; he later attended a boarding school in Switzerland. He studied chemistry in Europe and did research on amylases, enzymes that break down starches—much like Takamine's diastase. With limited academic opportunities in post–World War II Europe, Fischer moved to the United States, arriving at the University of Washington about five years after Krebs.

Krebs said in his Nobel Lecture that "the original studies of the Coris, which led to the finding of [glycogen] phosphorylase itself, grew out of the longstanding interest of these investigators on the role of [adrenaline] in the regulation of glycogen metabolism."

How Does Adrenaline Increase the Accumulation
of cAMP in Cells?

The acceptance of Sutherland's cAMP second-messenger theory focused attention on a new problem: how did adrenaline and other hormones activate adenylyl cyclase? Adenylyl cyclase resides on the inside of the plasma membranes that surround the contents of each cell. The simplest hypothesis was that adrenaline receptors and adenylyl cyclase constituted one molecule, with a part sticking out of the cell to spot adrenaline and another component inside the cell to make cAMP. Perhaps the binding of adrenaline to the receptor part of the molecule led to some sort of twist that activated the enzyme. Sutherland favored the working hypothesis that the receptors were an integral part of the adenylyl cyclase system, either one molecule or, alternatively, two closely associated but distinct subunits.

The determination of how adrenaline activated adenylyl cyclase proved quite complicated but fundamental to the actions of many hormones and drugs. Subsequent to Sutherland's pioneering studies, innovative research requiring new methods demonstrated that adrenaline receptors (β receptors in particular) do not activate adenylyl cyclase directly but rather require a previously unknown protein to serve as the critical intermediary between the receptors and adenylyl cyclase.

Martin Rodbell, who played a pivotal role in solving this problem, attributed some of his interest in second messengers in hormone action to a stimulating lecture by Earl Sutherland at NIH circa 1966. Rodbell received his PhD in biochemistry from the University of Washington in 1954. After a postdoctoral fellowship, he took a position at the NIH, focusing on the biochemistry of fat and cholesterol metabolism. Several years before Sutherland's transformative talk, Rodbell perfected techniques for digesting fat to obtain isolated fat cells in order to continue his research on lipoproteins. Bernardo Houssay challenged Rodbell about the viability of his isolated fat cells. A few days later, Rodbell showed Houssay results demonstrating that these fat cells responded to insulin. Houssay expressed excitement about Rodbell's capacity to study hormone action in isolated cells; his enthusiasm considerably encouraged Rodbell, who felt, "Practically overnight I had become an endocrinologist."

Lutz Birnbaumer, who arrived in Rodbell's laboratory in Bethesda in the late 1960s, recounted a typical day in the Rodbell lab at that time: remove the special fat pads from near the testicles of twenty rats, digest the fat pads to obtain isolated fat cells, and measure adenylyl cyclase responses to hormones such as insulin, adrenaline, and glucagon. Rodbell's laboratory had considerable interest in measuring the capacity of glucagon, a peptide hormone, to bind to its receptors and activate adenylyl cyclase. ATP, the starting material for the formation of cAMP, is an essential ingredient in adenylyl cyclase experiments. A control experiment that added ATP to the receptor experiments unexpectedly demonstrated that ATP apparently modified the interaction between glucagon and its receptor. Figuring out the basis for this strange result led to a major discovery. Rodbell's laboratory used ATP that had been purified from animal tissues by a commercial firm; it turned out that these preparations were contaminated by small amounts of guanosine triphosphate (GTP).[13] Further experiments demonstrated that the effects on receptors were due to the GTP rather than ATP itself. Rodbell's group went on to demonstrate that tiny amounts of GTP were essential for the activation of adenylyl cyclase by glucagon, adrenaline, and other hormones.

Rodbell conceptualized hormone action as a flow of information.[14] He credited this viewpoint to his experience as a radioman in the navy during World War II, when he would spend all day listening to Morse code. In the broadest sense, the flow of information with adrenaline starts with the brain perceiving a threat, triggering release of adrenaline, which carries the message to target cells through-out the body. Specific receptors on these cells recognize adrenaline, which passes the signal to adenylyl cyclase, which increases cAMP production, which in turn activates PKA, causing the cell's physiolog-ical response to the stressful situation. These processes within cells are known as signal transduction pathways. Rodbell used pictorial models to envision how hormones activated adenylyl cyclase; he pro-posed that GTP bound to an unknown transduction component that interacted with the receptors and transmitted the hormone sig-nal to adenylyl cyclase.

Further insight about the role of GTP followed from work by Dan Cassel and Zvi Selinger in Israel, who demonstrated that activation of β receptors enhanced the breakdown of GTP to guanosine di-phosphate (GDP). Moreover, inhibiting the breakdown of GTP markedly increased the capacity of adrenaline to activate adenylyl cyclase. These investigators proposed that adrenaline-mediated activa-tion of adenylyl cyclase involved the replacement of GDP with GTP somewhere in the system, and that breakdown of GTP to GDP turned off the system. The GTP transduction component suggested by Rod-bell turned out to be the on-off site where GTP bound and was me-tabolized to GDP.

Identification of the GTP binding sites—now called G proteins—depended on a key discovery that had no apparent connection to the role of GTP in signal transduction. This discovery owes much to Gordon Tomkins in San Francisco.[15] The general problem of how hormones activate cellular responses intrigued Tomkins. He applied a genetic perspective to his experiments by using s49 cells, which had the odd property of being killed by cAMP. Growing s49 cells with a chemical that mimicked cAMP killed almost all of them; rare mutant cells survived because they lacked PKA activity, and consequently did not respond to cAMP.

The s49 cells have β adrenergic receptors that increase cAMP accumulation in response to adrenaline. In the mid-1970s, Tomkins, with collaborators Henry Bourne and Philip Coffino at the University of California, San Francisco, demonstrated that stimulating these receptors with a drug related to adrenaline killed almost every s49 cell. However, some rare surviving mutants stayed alive because their cAMP concentrations did not increase. These investigators concluded that the mutant cells lacked the enzyme adenylyl cyclase, and they named them adenylyl cyclase minus cells, or cyc⁻ cells. Paul Insel in San Francisco, Alfred ("Al") Goodman Gilman in Virginia, and their colleagues demonstrated that these mutant cells had a normal number of β receptors, compatible with the hypothesis that β receptors and adenylyl cyclase came from distinct genes. Biochemical experiments in other laboratories demonstrated that β receptors and adenylyl cyclase molecules could be physically separated, which provided further evidence in support of this hypothesis.

Since β receptors and adenylyl cyclase likely constituted separate proteins, determining how the receptors activated the enzyme presented a major problem. Was there an unknown third component of the system that required GTP, or did GTP interact with β receptors or adenylyl cyclase directly? The remarkable answer came from Gilman's laboratory.

In his Nobel lecture, Gilman described Earl Sutherland's efforts to recruit him to Western Reserve University's combined MD-PhD program and the Department of Pharmacology. While the idea of spending seven years in Cleveland was not appealing, the excitement of cAMP, Sutherland, and the program more than compensated for the location. However, Gilman had some reluctance about joining a pharmacology department, as it would mean following in the footsteps of his highly esteemed pharmacologist father, Alfred Gilman. The senior Gilman is best remembered professionally for his authorship with

Louis Goodman of the definitive textbook of pharmacology—known affectionately by pharmacologists around the world as Goodman and Gilman.[16] The first edition appeared in 1941, the year Gilman was born. With his middle name, "Goodman," he became, according to the younger Gilman's colleague Michael Brown (who shared the 1985 Nobel Prize in Medicine with Joseph Goldstein), the only person ever named after a textbook.

Sutherland successfully reassured the young Al Gilman that pharmacology at Cleveland was really biochemistry with a purpose. However, when Gilman arrived in Cleveland in 1962, Sutherland was in the process of moving to Vanderbilt, so Gilman undertook his graduate work with Ted Rall, the investigator who had played a major role in the discovery of cAMP. As a graduate student, Gilman worked on cAMP responses in the thyroid gland. He undertook postdoctoral training at NIH in the laboratory of Marshall Nirenberg (who shared a Nobel Prize in Medicine in 1968 for work on the genetic code). Gilman studied the actions of adrenaline analogs on cAMP accumulation in isolated nerve cells. He also developed a novel, greatly simplified method to assay cAMP that opened research on cAMP to nonspecialized laboratories around the world; the article describing this method has been cited in more than five thousand publications. Gilman then moved to a faculty position at the University of Virginia, where he continued research on adrenaline signaling. After several years of hard work, the difficulties in making progress with the biochemistry of adenylyl cyclase prompted Gilman to express his frustration in a review article:

> The past few years have witnessed an extraordinary amount of information gathering on this subject [adenylyl cyclase], and we review it primarily with the goal of the clarification of our own thinking. It should be noted that, with the cerebral equipment available to us, this goal is left almost entirely unachieved. If this prolix introduction sounds pessimis-

tic, it is intentional. While the past two decades have demonstrated the presence of adenylate cyclase [former name of adenylyl cyclase] in a myriad of cell types, remarkably few fundamental observations have been made on this enzyme system since its elementary properties were described by Sutherland and Rall and their associates.[17]

In the article, Gilman predicated that the best approach going forward would involve the deep analysis of a small number of model systems rather than superficial characterizations of the properties of the enzyme in many cells and species. Indeed, within several years, a powerful genetic approach using mutant s49 cells provided a major breakthrough in his laboratory. In the late 1970s, Elliott Ross in Gilman's laboratory began seminal experiments with the cyc⁻ cells.[18] A general aim in these experiments involved the reconstitution of the capacity of adrenaline to activate adenylyl cyclase in plasma membranes from cyc⁻ cells; adding extracts containing adenylyl cyclase from other cells would provide the enzyme to the deficient cyc⁻ cells. These technically very challenging experiments finally worked: adrenaline activated adenylyl cyclase in reconstituted cyc⁻ cell membranes. However, the explanation for the results contained a dramatic and unexpected twist.

Careful scientists include many controls in their experiments to shield themselves from being misled. Ross and Gilman included a critical control experiment in which they destroyed the adenylyl cyclase enzyme in the donor extracts before combining them with the membranes from cyc⁻ cells, expecting that these extracts would be ineffective. However, these extracts nonetheless restored adrenaline's capacity to increase cAMP synthesis in the cyc⁻ cells' membranes. Figuring out the explanation for this startling result transformed the course of their work. Cyc⁻ cells had been, in retrospect, misnamed; they contained hidden, essentially silent adenylyl cyclase. The mutant cells actually lacked a previously unknown component required for robust activation of adenylyl cyclase. The donor extracts provided this missing component to the membranes from cyc⁻ cells.

Gilman and his colleagues then purified this mysterious missing component, elegantly discovering a novel GTP binding protein, now called G_s. Instead of missing adenylyl cyclase, the mutant cyc⁻ cells lacked G_s. G_s represented the biochemical reality of the GTP-dependent transducer proposed by Rodbell. A fully compelling explanation required putting purified protein components together to build a working system, so Gilman's laboratory assembled purified β receptors, G_s, and adenylyl cyclase and demonstrated that these proteins are sufficient to make a functional system that made cAMP in response to adrenaline-like drugs. Solving how adrenaline and other hormones activated adenylyl cyclase led to a Nobel Prize in Medicine in 1994 for Rodbell and Gilman.[19]

Subsequent research demonstrated that each G protein is made up of an α subunit and a $\beta\gamma$ subunit, with multiple genes encoding each of the α, β, and γ subunit proteins. Individual G proteins are named based on their α subunit; for example, G_s has an α_s subunit. In the resting state, the α_s and $\beta\gamma$ subunits are joined together with a GDP molecule attached to α_s. Activation of β adrenergic receptors by adrenaline leads to the replacement of GDP with GTP on α_s, and the separation of the α_s and $\beta\gamma$ units. The receptor's job is to promote the exchange of GTP for GDP. The α_s-GTP complex then combines with adenylyl cyclase, increasing the rate of cAMP synthesis. Subsequently, the α_s subunit metabolizes the GTP to GDP, turning off its own capacity to activate adenylyl cyclase, and then recombines with a $\beta\gamma$ subunit, returning G_s to the resting state. The system works as a switch, turning on and off depending on the presence of adrenaline. (See Figure 7-2.)

Abnormalities in G proteins, either inherited or acquired, can cause disease. To take one example, the bacterium that causes cholera, a

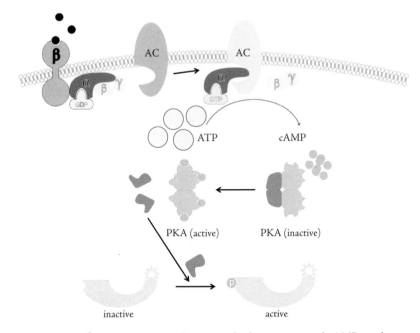

Figure 7-2. β receptors activate G proteins, leading to increased cAMP production, PKA activation, and phosphorylation of target proteins. The next major advances in signaling mediated by β receptors demonstrated that these receptors and adenylyl cyclase are distinct molecules that interact indirectly via activation of G proteins made up of two distinct parts, the α and the $\beta\gamma$ subunits. Activation of β receptors by adrenaline (●) promotes the association of the receptors with inactive G proteins. This interaction leads to the dissociation of GDP and association of GTP on the α subunit; the now free α subunits activate adenylyl cyclase (AC). cAMP binds to regulatory sites on inactive PKA, which leads to the dissociation of the now active catalytic subunits of PKA, which in turn phosphorylate target proteins.

life-threatening disease with profuse diarrhea, releases a toxin in the intestines that enters cells lining the bowels and dramatically increases cAMP in those cells. The sharp rise in cAMP causes the cells to secrete enormous amounts of fluid into the bowels, leading to massive diarrhea. The toxin works by adding a small chemical group to the α

subunit of G_s. This change in the α_s subunit inhibits its capacity to break down bound GTP to GDP, leaving it irreversibly activated. As another example, whooping cough is caused by infection with the organism *Bordetella pertussis*. This organism secretes multiple toxins; one of them contributes to the disease by inactivating a different G protein, called G_i. Inherited mutations in the α_s subunit can lead to abnormal cellular responses to hormones and the development of tumors.

⌐⠸⌐

The discovery of receptors that sit quietly in cell membranes awaiting the arrival of adrenaline initially involved pharmacological research that occurred in parallel with other great advances in signaling. In the next chapter we will take a close look at adrenaline receptors and how they spurred research that led to the invention of novel drugs.

EIGHT

Lock and Key:
Receptors for Adrenaline

To most modern pharmacologists the receptor is like a beautiful
but remote lady. . . . He cannot, however, truly claim ever
to have seen her, although one day he may do so.

D. K. DE JONGH, 1964

Adrenaline operates only on cells that have adrenergic receptors; by
sliding into these intricate structures like a key, adrenaline unlocks
each receptor's capacity to set in motion cascades of signals. The hypothesis that drugs interact with specific receptors emerged at the
end of the nineteenth century. However, skepticism about the existence of receptors gave way only relatively recently, under the weight
of increasingly powerful experimental evidence. Research over the
past hundred years addressing adrenaline action has played a pivotal
role in recognizing that drug and hormone receptors are concrete
molecules that have specific biochemical effects in cells.

The Origin of the Idea of Receptors

The hypothesis that drugs initiate their actions by combining with
specific receptors owes much to the work of Paul Ehrlich and John
Langley around the turn of the twentieth century. Ehrlich developed
the concept of receptors to explain workings of the immune system, and
later he extended the idea to drugs. Ehrlich's idea was epitomized by
the maxim *Corpora non agunt nisi fixate*—chemical agents do not act

unless they are bound. In the ideal case, the binding between a drug and target of action would be so specific that no other process would be affected. Such a drug would be like a magic bullet—for example, killing bacteria without affecting the patient in any other way.[1]

Langley demonstrated that atropine blocked the capacity of pilocarpine to stimulate the flow of saliva. He suggested that these chemicals opposed each other by interacting with a common structure involved in the secretion of saliva. Langley later demonstrated that curare blocked nicotine-stimulated skeletal muscle contraction; he proposed that nicotine and curare interacted with the same "receptive substance." In modern terms, nicotine is an agonist that activates receptors involved in muscle contraction; curare is an antagonist that occupies the same receptors without activating them. In the presence of high concentrations of curare, nicotine cannot beat out curare for a place on the receptors, so contraction does not occur.

At this point in time, however, the concept of receptors did not yet have a compelling scientific foundation. Receptors were invisible and could not be measured directly. Available chemical theory could not readily account for the selective binding of drugs to receptors. Many pharmacologists believed that drug action could be explained by each drug's physical properties, such as solubility. For example, Walther Straub postulated that drug concentration differences between the outside and inside of cells accounted for drug effects. Conclusive demonstration of the physical reality of receptor molecules progressed only slowly over the next seventy years, with research on potential adrenaline receptors playing a leading role. Nonetheless, over the next several decades more than a few investigators made interesting theoretical inferences about the properties of the alleged receptors.

In 1909, A. V. Hill, an undergraduate student working with Langley, mathematically described relationships between drug concentrations and their effects.[2] Alfred Clark and John Gaddum independently extended Hill's results in 1926; their work developed important ideas that applied the concept of receptors to quantitatively analyze and predict drug effects.[3] Eleven years later, Gaddum derived an important equation that described competition between agonists and antagonists for the same receptors. Ten years after that, Heinz Schild furthered this work with additional quantitative methods for determining the attraction (affinity) of antagonist drugs for receptors.[4]

The broad success of abstract receptor theory in predicting drug effects strengthened the belief in the existence of receptors. However, no receptor had been nailed down physically, leaving considerable room for skepticism. The theory of receptors was further elaborated by E. J. Ariëns in Holland and R. P. Stephenson in Scotland, who discovered in the mid-1950s that some drugs had effects intermediate between agonists and antagonists: these drugs became known as partial agonists. Mark Nickerson demonstrated that activation of only a fraction of the adrenaline receptors in blood vessels provided a sufficient stimulus to fully contract the arteries. In other words, there were extra or "spare" receptors; these extra receptors enabled blood vessels to respond to very low concentrations of adrenaline.[5] Nonetheless, receptors remained abstract entities, something that many mainstream scientists thought were convenient fictions.

Dual Receptors for Adrenaline

In 1948, Raymond Ahlquist pushed the notion of receptors closer to accepted reality, albeit with a slow takeoff. Ahlquist came up with a new formulation for evaluating the responses mediated by adrenaline. He proposed a revolutionary concept: that adrenaline activated two distinct receptors, which he called α and β adrenergic receptors.

After receiving his doctorate in pharmacology in 1940, Ahlquist worked in South Dakota on a project aimed at domestically growing the plant *Ephedra sinica*, the natural source of ephedrine (discussed in Chapter 9) because supplies of this adrenaline-like drug were threatened by Japanese military action in China. In 1944, Ahlquist joined the faculty of the Medical College of Georgia. Raymond Woodbury, the chair of pharmacology, sought drugs that relaxed the uterus as a potential therapy for painful menstruation. His interests stimulated Ahlquist to test adrenaline, noradrenaline, and three closely related analogs on the uterus and multiple other animal tissues. Ahlquist ranked the drugs from largest to smallest effect in each of the tissues and found exactly two different drug sequences for the effects of the five drugs in all these organs.

He imaginatively conjectured that his data could be explained by the existence of two different receptors for adrenaline, with different affinities for the test drugs, as represented by the two drug sequences. He gave these hypothetical receptors the noncommittal names α and β adrenergic receptors. The test drug isoproterenol had the most value in differentiating between these receptors: isoproterenol powerfully activated β receptors but had essentially no activity at α receptors. In Ahlquist's formulation, when adrenaline combined with β receptors, the heart was stimulated but smooth muscle relaxed. Fifty years earlier, Langley had suggested that adrenaline caused smooth muscle contraction in some tissues and relaxation in others by activating excitatory and inhibitory receptors, respectively. In Ahlquist's framework, a single β receptor could mediate excitatory or inhibitory responses in different tissues. We now understand the molecular basis that explains how Ahlquist was correct. The binding of adrenaline to β receptors activates cAMP accumulation both in the heart and in smooth muscle. In the heart, cAMP excites the rate and force of contraction, whereas in smooth muscle cAMP inhibits contraction. In other words, activated β receptors can either excite or inhibit physiological responses depending on the properties of the tissue.

Ahlquist's proposal seriously conflicted with a hypothesis promulgated by the late Walter Cannon, namely, that the interactions of excitatory and inhibitory forms of adrenaline activated a single adrenergic receptor, leading to excitatory and inhibitory effects, respectively. Ahlquist failed to get his manuscript accepted for publication in the leading pharmacology journal in the United States; he felt deep and bitter resentment about this rejection, attributing the outcome to his criticism of the ideas of the very famous Cannon. Ahlquist ended up publishing his paper in 1948 in a physiology journal, an outcome possibly facilitated by a friendly editor. Though this paper ultimately became famous, referenced in more than 2,500 subsequent scientific articles, for a decade after its publication few investigators paid any attention to the so-called α and β adrenergic receptors.

Ahlquist did not follow up with important new experimental findings involving α and β adrenergic receptors. He did use this theoretical framework in a 1954 textbook chapter to simplify the description of autonomic nervous system pharmacology. Presenting fundamental information using an unaccepted theory in a student textbook is most unusual; these books typically present only the most secure and well-accepted material. This otherwise unexceptional chapter had a remarkable and unanticipated impact on the future of pharmacology and medicine in the second half of the twentieth century, as we shall soon see.

Blocking Adrenaline at α Receptors

In 1906, Dale reported that an ergot extract called ergotoxine blocked some of the stimulatory effects of adrenaline and the sympathetic nervous system. Ergotoxine attenuated adrenaline-mediated smooth muscle contraction but not adrenaline's capacity to increase heart rate.[6] In retrospect, we now understand that Dale's ergot preparation contained an α adrenergic antagonist that left β receptors untouched. Dale only casually mentioned Langley's concept of receptors in interpreting his results. Dale did not endorse the existence of receptors with any enthusiasm over the succeeding decades, perhaps viewing the concept as unnecessarily speculative.

In the 1920s, chemists made progress in synthesizing substances that attenuated the capacity of adrenaline to cause vasoconstriction. Daniel Bovet, working with the French medicinal chemist Ernest Fourneau, evaluated the capacity of dozens of compounds to block or mimic adrenaline-mediated increases in blood pressure. Many of these chemicals blocked what we now call α receptors.

Bovet, born in Switzerland, received his doctorate in 1929. He worked at the Pasteur Institute in Paris for the next twenty years and later assumed a senior academic appointment in Italy.[7] When he received the Nobel Prize in Medicine in 1957 for his work in medicinal chemistry, especially for inventing drugs that block the effects of histamine, adrenaline, and acetylcholine, the presentation speech by a member of the Royal Caroline Institute was remarkably prophetic: "Sympatholytic compounds have not yet found any application in general medicine. The future will tell whether the hopes placed in them will be fulfilled and whether they will be of value in the treatment of hypertension and other vascular conditions for which we think a reduction in nervous control would be desirable."[8]

β Adrenergic Receptor Antagonists: Transformative Drugs

Ahlquist's framework quietly suggested the theoretical possibility that novel chemicals might specifically block β receptors. In 1958, with quite different intentions in mind, C. E. Powell and I. H. Slater at Eli Lilly and Company in Indianapolis found that a new chemical, dichloroisoproterenol (DCI), had some unusual pharmacological actions: it antagonized adrenaline-mediated dilation of airways in the lungs without attenuating adrenaline-mediated constriction of blood vessels. In describing these and other actions of DCI, the scientists suggested that DCI combined with adrenergic inhibitory receptor sites, consistent with Langley's views about adrenaline receptors from a half century before; they did not refer to Ahlquist's ideas about α and β adrenergic receptors published ten years earlier.

About a year later, Neil Moran in Atlanta published elegant and comprehensive pharmacological experiments with DCI that he had obtained from Eli Lilly. Moran, however, explicitly used Ahlquist's formulation of α and β receptors in interpreting his results, suggesting that DCI was the first known example of a β receptor antagonist.[9] Ironically, Moran's paper was published in the pharmacology journal that had rejected Ahlquist's original paper. Moran's results with DCI and their interpretation considerably advanced the stature of Ahlquist's formulation by demonstrating the existence of a β adrenergic receptor antagonist. A few years later, DCI provided an important launching pad for the research of James Black, leading to the invention of β receptor antagonist drugs for the treatment of angina in patients with coronary artery disease.

James Black, born into a coal mining family in Scotland, graduated in medicine from the University of St. Andrews in 1946. After a year in research that enhanced his interests in cardiovascular and gastrointestinal physiology, Black moved to Singapore as a lecturer at the King Edward VII College of Medicine (now part of the National University of Singapore), where he mainly taught physiology, and also did some research on the control of blood flow. After returning to the United Kingdom three years later, Black soon headed the new physiology department at the University of Glasgow Veterinary School.

Black decided to focus his own research on drug therapy of angina, a clinical syndrome first described in the late eighteenth century.

With physical activity adrenaline and the sympathetic nervous system stimulate the heart to beat harder and faster to provide more oxygen and nutrient-rich blood to exercising skeletal muscles. The added work of the heart increases its own requirements for oxygen. Normal coronary arteries dilate during exercise, increasing blood flow to the heart muscle to meet that need. However, partially occluded atherosclerotic coronary arteries fail to dilate adequately; inadequate coronary blood flow leads to oxygen debt in the heart muscle, giving rise to the disagreeable sensation or pain in the chest called angina. Angina is relieved by rest, which allows the heart muscle's oxygen requirements to decrease.[10]

In the 1950s, many investigators interested in developing new medical treatments for angina sought drugs that dilated coronary arteries. However, Black had reservations about the potential success of this strategy, as he doubted that drugs could effectively dilate stiff, diseased coronary arteries. Black wondered if a drug that decreased the heart muscle's requirement for oxygen during exercise or stress would alleviate angina. He reasoned that preventing the rise in heart rate that occurs with exercise might accomplish this goal.[11]

Black fortuitously chose Victor Drill's 1954 *Pharmacology in Medicine* textbook to learn more pharmacology. Reading Ahlquist's material exposed him to the concept of α and β adrenergic receptors. Noting that β adrenergic receptor activation stimulated the heart, Black decided that he wanted to invent a β adrenergic receptor antagonist to treat angina.

Any serious plan to synthesize and evaluate potential β adrenergic receptor antagonists required considerable financial and technical backing. Black approached Imperial Chemical Industries (ICI) seeking funding for this ambitious and risky project. ICI was a major firm, founded in 1926 through the merger of four large chemical companies, including Alfred Nobel's dynamite business.[12] Black's proposal intrigued representatives from ICI, who made a startling counterproposal: that Black move to a position in ICI's laboratories in England to pursue his bold hypothesis. In 1958, leaving a university position for the pharmaceutical industry represented a significant career transformation for an academic scientist (much as it had for Henry Dale more than half a century earlier).

In the effort to synthesize a β adrenergic receptor antagonist at ICI, Black and his collaborator, chemist John Stephenson, started with analogues of the agonist isoproterenol, a drug highly selective in activating β receptors compared to α receptors, as noted by Ahlquist. Chemical modifications aimed at transforming this powerful agonist into a β receptor antagonist proved unsuccessful. However, the recently published work on DCI provided new ideas for making β adrenergic receptor antagonists.[13] Stephenson replaced the chloride atoms in DCI with a different chemical group, giving rise to a novel chemical called pronethalol. Pharmacological experiments demonstrated that pronethalol was a β adrenergic receptor antagonist.

ICI moved pronethalol forward in clinical trials for treating patients with angina; the exciting outcome was that this drug actually improved angina, as predicted by the bold concept Black had developed years earlier. Unfortunately, pronethalol caused impaired coordination and vomiting; these adverse effects proved unrelated to its blockade of β receptors, leaving the door open to develop another β receptor antagonist without these adverse effects. Toxicology studies also demonstrated that pronethalol caused tumors in mice, making it even less attractive. Black's group then synthesized propranolol, which improved on pronethalol. Propranolol is a very potent β receptor antagonist that did not demonstrate troublesome problems in toxicology studies; in patients propranolol had only relatively minor nonspecific adverse effects.[14] Propranolol proved highly effective in relieving angina in early clinical trials.

In 1964, with β adrenergic receptor antagonists for angina now on firm footing, Black decided to seek a new challenge in early drug development, preferring not to get involved in the inevitable development work required to launch and market a new drug. He moved to the pharmaceutical firm Smith, Kline & French, where he invented an H_2 receptor antagonist. The drug blocked histamine's capacity to stimulate acid secretion in the stomach and proved effective in the treatment of peptic ulcer disease.[15] Black, together with his colleagues in medicinal chemistry, had emasculated two powerful agonists, adrenaline and histamine. For his work developing β receptor and H_2 receptor antagonists, Black shared half the Nobel Prize in Medicine in 1988. Upon learning that he had received the award, Black joked, "I wish I had some of my beta blockers handy." He wryly noted later that receiving the prize was "double-edged" in that it provided "limitless opportunities to waste my time on small bits of personal publicity." However, for the rest of his life he actually focused on exciting research programs, once commenting, "Pharmacologists are addicted to receptors, to their subdivision and classification, to their explanatory properties, and to their use as targets and templates for designing new drugs."

The synthesis of additional novel drugs suggested that Ahlquist's schema of α and β adrenergic receptors needed refinement. Black's group invented several compounds that differed in their capacity to block β adrenergic receptors in the heart compared to β receptors in arteries. In 1967, Lands and collaborators formally described two distinct subtypes of β adrenergic receptors: β_1 receptors mediate adrenaline's effects in the heart, whereas β_2 receptors mediate dilation of smooth muscle in arteries and airways. About twenty-five years later, β_3 receptors came to light in fat. Moreover, additional studies with drugs not available to Ahlquist indicated that not all α receptors are created equal. By 1980, pharmacologists recognized two major α receptors, α_1 and α_2 receptors. The implications of these results for drug discovery are described in Chapter 9.

Adrenergic Receptors: Ghosts of Departed Quantities

In the eighteenth century, Bishop Berkeley criticized the mathematical foundation of calculus, which then was based on infinitesimals, individual quantities so small they could not be measured; Berkeley called infinitesimals "the ghosts of departed quantities." Something similar might have been said of adrenergic receptors until the 1970s: while the abstract conception of α and β adrenergic receptors provided the framework for the development of β adrenergic receptor antagonist drugs, the receptors themselves remained invisible and had not yet been measured directly. Even Ahlquist himself expressed many doubts about the true nature of receptors.

In the 1970s, scientists using radioactive drugs succeeded in directly determining the number of α and β adrenergic receptors in tissues, in a sense visualizing them for the first time. The idea is relatively straightforward: select a drug that is very potent in interacting with an adrenergic receptor, tag the drug with a radioactive atom, and use the radioactive drug to label receptors. In practice, cells have few receptors and drugs interact nonspecifically with many irrelevant sites. Consequently, differentiating between the small signals generated by genuine receptors and background noise proved difficult. Success finally came in the early part of that decade. Robert Lefkowitz at Duke University led efforts using radioactively tagged β receptor antagonist drugs to measure β receptors in tissue extracts. Within several years, radiolabeled drug assays capable of measuring α_1 and α_2 receptors also became widely available. With these advances, adrenergic receptors finally became more tangibly real for many pharma-cologists. Moreover, the radiolabeled drugs opened the door to puri-fying the receptors from tissue extracts, since the receptors could now be radioactively tagged.

Purifying adrenergic receptors proved exceedingly difficult since each cell has relatively few copies of these receptors. Consequently, extracts from tissues required enrichment by about 100,000-fold to isolate receptors freed from the other materials in cells. In the early 1980s, after intensive and difficult work over many years, Lefkowitz and colleagues isolated microgram quantities of pure β receptors. Critical experiments demonstrated that the material represented a functional β adrenergic receptor: adrenaline-like drugs stimulated cAMP production when the putative β receptors were combined with adenylyl cyclase and G_s. After almost a century of skepticism, the physical reality of adrenaline receptors had finally been established.

In 1986, Lefkowitz and collaborators obtained the DNA sequence encoding the β_2 receptor, representing the first cloning of a G-protein-coupled hormone or drug receptor gene. Remarkably, the amino acid sequence of the β_2 receptor proved very similar to the known struc-ture of rhodopsin, the eye protein responsible for detecting pho-tons.[16] This astonishing result provided the first hint that G-protein-coupled receptors for light, adrenergic receptors, and receptors for many other hormones and drugs had previously unrecognized struc-tural similarities.

Isolation of the β_2 receptor gene, coupled with additional progress in receptor purification and DNA-based technologies, made it possi-ble to identify the genes for all the adrenergic receptors. Nine differ-ent gene sequences for adrenergic receptors emerged: three different β receptor genes (β_1, β_2, and β_3), three α_1 receptor genes (α_{1A}, α_{1B}, and α_{1D}), and three α_2 genes (α_{2A}, α_{2B}, and α_{2C}).[17] The list of nine adrenergic receptors is likely definitive.

Subsequent research confirmed that all the adrenergic receptors have similar, complicated shapes; each receptor protein passes through the cell surface membranes seven times. Receptors with this overall configuration are called seven-transmembrane domain (7TM) receptors or serpentine receptors because of their overall appearance; because these receptors couple to a variety of G proteins, they are also called G-protein-coupled receptors (sometimes abbreviated as GPCR). In 2007, Brian Kobilka, a former research fellow with Lefkowitz, and colleagues determined the crystal structure of the human β_2 receptor, providing tools for a new understanding of physical changes in the receptor as it binds adrenaline.[18] The Nobel Prize in Chemistry in 2012 was shared by Lefkowitz and Kobilka for their studies of G-protein-coupled receptors.[19]

The genes encoding the 7TM receptors constitute one of the larger families of related genes in the human genome. The diversity of G-protein-coupled receptors is enormous. Members of 7TM family include receptors for morphine and histamine, receptors that recognize single atoms such as calcium, receptors for large pituitary hormones, and the receptors in our nose that allow us to distinguish thousands of different smells.[20] Remarkably, the HIV virus, responsible for the AIDS epidemic, gets into cells by exploiting a 7TM cell surface receptor. More than 20 percent of marketed drugs achieve their effects by interacting with a G-protein-coupled receptor.[21] Molecular biologists have identified more than three hundred genes for receptors that share the overall shape of the adrenergic receptors in the human genome (not counting the thousands of receptors for smell and taste). Many of these 7TM receptors lack an obvious hormone or neurotransmitter capable of activating them; these are the so-called orphan receptors. "Deorphanizing" an orphan receptor involves identifying the natural substance that activates it. This prob-

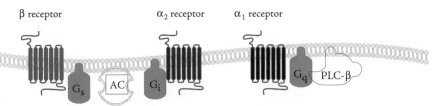

Figure 8-1. The three major classes of adrenergic receptors. α_1, α_2, and β receptors, have the same general structure but differ importantly in amino acid sequences. These receptors stimulate distinct downstream signaling pathways by coupling to different G proteins, called G_s, G_i, and G_q. G_s and G_i respectively increase and decrease the rate of synthesis of cAMP by adenylyl cyclase (AC). G_q activates the enzyme phospholipase C_β, which cleaves a lipid molecule, liberating two second messengers that activate protein kinases and change the intracellular concentration of calcium ions.

lem represents the inversion of the traditional historical challenge of identifying receptors for known substances: in this case the receptor is at hand, but the substance that it interacts with must be tracked down by reverse pharmacology. Lefkowitz and collaborators deorphaned the first orphan receptor by identifying a DNA sequence encoding a 7TM receptor that they later determined was activated by serotonin, a neurotransmitter important in the brain. More than a hundred 7TM receptors remain orphans, however. Deeper understanding of the function of these receptors offers the possibility of identifying new targets for drugs that regulate their activity.

The β_1, β_2, and β_3 receptors each couple with G_s to activate adenylyl cyclase, increasing cAMP accumulation. On the other hand, each of the three α_2 receptor subtypes inhibits adenylyl cyclase activity by coupling to the G proteins G_i and G_o, lowering the accumulation of cAMP. The fall in cAMP accumulation turns down the activity of protein kinase A. The α_1 receptors function largely separately from adenylyl cyclase. They couple with G_q and G_{11}, which activate the enzyme phospholipase C_β, leading ultimately to increased intracellular calcium and activation of protein kinases by several pathways.

Cells have mechanisms that turn off signaling when adrenaline receptors are too strongly activated for extended intervals. These adaptations decrease the chance that overstimulation might damage cells. An important feedback mechanism involves phosphorylation of these receptors by a family of specialized protein kinases. After these receptors are phosphorylated, specialized small proteins attach to the receptors; these proteins decrease the capacity of the receptors to activate G proteins.[22]

Why Are There So Many Adrenaline Receptors?

From an evolutionary viewpoint, G-protein-coupled receptors have been around for hundreds of millions of years. Primitive life-forms such as slime molds have them, as do some bacteria that use a form of rhodopsin to capture energy from sunlight. Genetic diversity increases when genes duplicate themselves and then mutate and evolve separately. Some of these divergent genes code for receptors with favorable biological characteristics that are perpetuated as a species evolves. Gene duplication has led to new classes of receptors and subtypes within a class. The α_1, α_2, and β receptors emerged and further evolved in this manner. The expression of multiple receptors for adrenaline provides opportunities for subtler responses as well as for magnifying the potential effects of receptor activation. A further level of complexity has been revealed by the sequencing of the adrenergic receptor genes in large numbers of people around the world. We now know that there are many polymorphisms in the adrenergic receptor genes. Polymorphisms are relatively common variations in the DNA sequence for a gene within a population; for example, blood group polymorphisms in DNA give rise to the various blood groups. Polymorphisms in the adrenaline receptor genes have been associated with propensity to disease and have been used to predict responses to drugs; for example, the effect of β adrenergic agonists in asthmatics may be modified by the DNA sequence of the β_2 receptor gene in some of these patients. Many efforts are under way to exploit this type of information as part of a large effort that has been dubbed "personalized medicine."

The recognition of the three major types of adrenaline receptors (β, α_1, and α_2), each with three subtypes, provides a foundation for understanding and predicting the effects of the many drugs that activate or inhibit responses mediated by adrenaline or the sympathetic nervous system. In addition, this framework provides an opportunity to develop novel drugs with more selective actions. These topics will be pursued in the next chapter.

NINE

New Drugs from Old Molecules

The most fruitful basis for the discovery of a new drug is
to start with an old drug.

ATTRIBUTED TO JAMES BLACK

Early twentieth-century chemists seized the opportunity to synthesize structurally related analogs of adrenaline; pharmacologists evaluated these new substances in elaborate, time-consuming animal experiments. Some of these compounds partially reproduced or inhibited the actions of adrenaline. A number of interesting drugs emerged, providing either novel therapies for disease or serving as useful pharmacological tools that helped unravel the fundamental biology of the autonomic nervous system. Throughout the first half of the twentieth century, limited understanding of how adrenaline activated cellular responses hindered the invention of drugs with highly selective actions. In the second half of the century, Ahlquist's framework of α and β adrenergic receptors provided a stronger conceptual framework to seek adrenaline analogs with clinically desirable characteristics.

The finer classification of adrenergic receptors into three major classes—α_1, α_2, and β receptors, with their various subtypes—has provided an even firmer underpinning in the search for greater pharmacological precision. Chemists have developed powerful methods that facilitate the construction of libraries of thousands of analogs; pharmacologists have kept pace by implementing automated test-tube assays that can determine interactions with each of the adrenergic receptors with relatively little effort.[1] These advances have greatly speeded the rate at which chemicals can be synthesized and evaluated, although they do not necessarily provide a more effective approach to discovering important new drugs.

This chapter describes the development of important drugs related to adrenaline, starting with the naturally occurring substances tyramine and ephedrine, and then moving to chemically synthesized drugs with increasing specificity in their effects. Many of these drugs play central roles in the therapy of serious diseases in the twenty-first century. Their stories emphasize that successful drug invention and development often involve a fascinating interplay of fundamental knowledge of both pharmacology and disease, sometimes combined with astute clinical observations to recognize unanticipated valuable effects.[2]

Tyramine: The Unusual Origin of a Sympathomimetic Drug

Tyramine unexpectedly became a remarkable tool in unraveling intricate and important biological mechanisms in the autonomic nervous system, illustrating how apparently contradictory or anomalous observations can sometimes contribute to major discoveries.

In 1906, George Barger and Henry Dale at Burroughs Wellcome learned of the French discovery that extracts of putrid meat raised blood pressure. With his curiosity aroused, Barger allowed several pounds of first-rate beefsteak to putrefy and then experimentally confirmed the French scientists' results. Barger, Dale, and their colleagues scaled up the experiments using putrefied horse meat and identified tyramine as the most active blood-pressure-raising substance in the extracts. Tyramine and adrenaline had broadly similar actions, although injections of tyramine raised blood pressure more slowly but the effects lasted longer. Burroughs Wellcome marketed tyramine as one of the first sympathomimetic drugs used in patients.

Walter Dixon and Frank Taylor discovered one year later that extracts from placentas raised blood pressure. Otto Rosenheim in London decided to identify the active substance in the placenta; however, he could not initially reproduce their results. He worked out that the blood-pressure-raising activity was present only in putrefied extracts contaminated by bacteria; bacteria feeding on the placental extracts apparently produced active substances. Rosenheim succeeded in isolating from decaying placentas several substances that raised blood pressure. After learning about the work of Barger and Dale, Rosenheim determined that tyramine was also the active substance in putrid placentas.

We now know that tyramine is a breakdown product of the amino acid tyrosine. In the intestines, digestion of proteins liberates tyrosine; bacteria in the intestines transform some of the tyrosine into tyramine. In addition, large amounts of tyramine are already present in many fermented foods, especially some cheeses and wines. Tyramine is absorbed from the intestines; however, it is normally inactivated in the liver by the enzyme monoamine oxidase (MAO) before getting further into the body. As described in Chapter 6, some monoamine oxidase inhibitor drugs allow tyramine to pass through the liver into the general circulation, causing severe increases in blood pressure.

Tyramine never had a major role in medical therapeutics; however, sorting out some anomalies in the pharmacology of tyramine unexpectedly provided much deeper insight into the biology of sympathetic nerve endings. In 1927, investigators at Stanford University found that tiny amounts of cocaine blocked tyramine's capacity to raise blood pressure. Since cocaine reliably magnifies adrenaline's effects on blood pressure, this result was unexpected. Why did cocaine have such different effects on these two sympathomimetic drugs? The full explanation emerged more than thirty years later, after major discoveries provided a more thorough understanding of sympathetic neurons and the effects of cocaine on these cells.

After releasing noradrenaline, sympathetic neurons use a pump to sweep most of the released noradrenaline back inside these nerve endings. The noradrenaline reuptake pump has two major effects: the quick removal of noradrenaline from outside the neuron limits the duration and extent of its action on target cells, and the recycling of noradrenaline minimizes the sympathetic neuron's need to synthesize large amounts of this neurotransmitter. Remarkably, cocaine also binds to this specialized pump and blocks the transport of noradrenaline (or adrenaline) into the neuron. Inhibiting the reuptake of noradrenaline quickly raises the concentration of noradrenaline outside the nerve endings. This change leads to far greater stimulation of adrenergic receptors on target cells, especially α receptors in arteries, causing more marked constriction and a consequent rise in blood pressure.

Tyramine is also transported into sympathetic nerve endings by the noradrenaline reuptake pump, likely because tyramine has some structural features in common with noradrenaline. Once inside nerve endings, tyramine releases noradrenaline from sympathetic neurons; the released noradrenaline then activates α receptors in arteries, leading to the typical rise in blood pressure. By blocking the noradrenaline reuptake pump, cocaine prevents tyramine from entering these neurons and releasing noradrenaline. Additional experiments demonstrated that tyramine does not itself directly activate either α or β receptors. Consequently, in the presence of cocaine, tyramine has no

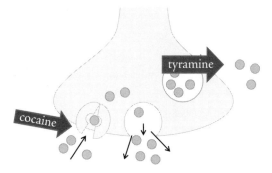

Figure 9-1. Tyramine and cocaine. Noradrenaline is taken up by a specific pump in sympathetic nerve endings and recycled for release as a neurotransmitter. Tyramine is taken up in these neurons by the same pump. Once inside the cells, tyramine promotes the release of noradrenaline. Cocaine blocks the pump that transports noradrenaline back into sympathetic nerve endings; this magnifies the effects of noradrenaline at the target cells' adrenergic receptors. Cocaine also blocks the entrance of tyramine into the neurons, which prevents tyramine from releasing noradrenaline, nullifying responses to tyramine.

hope of having a major effect on blood pressure.[3] The puzzle raised by the Stanford experiments ultimately had a very elegant solution.

The discovery of tyramine's indirect effects on blood pressure revealed a novel mechanism of drug action that proved relevant for other drugs. For example, powerful amphetamines also raise blood pressure largely by indirectly stimulating the release of noradrenaline from sympathetic neurons. In 1910, Dale coined the term *sympathomimetic* for chemicals that simulate the actions of the sympathetic nervous system; *sympatholytic* was later coined for drugs that blocked some of its actions. In 1958, J. Harold Burn divided sympathomimetic drugs into two groups: those that acted directly on adrenaline receptors and those that worked indirectly, primarily by releasing adrenaline or noradrenaline stored inside cells.

The Enigma of Clonidine: A Nasal Decongestant That Lowers Blood Pressure

Clonidine provides a dramatic example of how unanticipated actions of a new drug can lead to fundamental discoveries about human biology. Pharmaceutical companies have invented many α adrenergic agonists for indications ranging from life-threatening low blood pressure to opening blocked noses. For example, phenylephrine is a sympathomimetic drug, which raises blood pressure by constricting arteries. Phenylephrine also contracts smooth muscle in engorged nasal mucosa, shrinking its volume and thereby improving air movement through the nose in people with allergies or colds.[4] In the early 1960s, Boehringer Ingelheim invented clonidine, a sympathomimetic drug that ameliorated nasal congestion in animals. The secretary of a physician involved in clinical trials happened to have a stuffed nose at a propitious moment, and she became the first person treated with clonidine. The test dose proved far too high; while her congestion responded favorably, her blood pressure dropped precipitously, to the amazement and concern of coworkers.

Clonidine's capacity to improve nasal congestion and lower blood pressure represented an enigma: sympathomimetic drugs that caused smooth muscle contraction were expected to raise blood pressure. Going back to the drawing board with clonidine led to additional experiments that resolved the puzzle by revealing unexpected mechanisms involved in the control of blood pressure. Clonidine activates previously unknown α_2 receptors in the blood pressure control system in the brain. Activation of these receptors suppresses sympathetic nervous system activity throughout the body, leading to a fall in blood pressure; clonidine's effects in the brain are far greater than the much smaller constricting effects that clonidine has directly on arteries.[5] Consequently, an unanticipated adverse drug effect with clonidine led to the serendipitous discovery of a new biological system. Moreover, Boehringer Ingelheim turned the direction of development of clonidine around 180 degrees: new clinical trials transformed clonidine into a successful pill for lowering blood pressure in patients with hypertension. However, as might be predicted based on its capacity to suppress sympathetic nervous system activity throughout the body, postural hypotension is one of clonidine's adverse effects.

Ephedrine: A Born-Again Drug

The herb ephedra (*Ephedra sinica,* known in traditional Chinese medicine as ma huang) has been used to treat cough and stimulate the circulation for thousands of years. In 1887, Nagajosi Nagai in Tokyo reported the isolation of ephedrine from ephedra extracts.[6] A colleague demonstrated that ephedrine dilated the pupils; high doses were considered very toxic. Ephedrine had short-lived popularity in the form of drops to enlarge the pupils for eye examinations. In 1917, Japanese scientists demonstrated that ephedrine had sympathomimetic effects similar to adrenaline, including the capacity to relax bronchial smooth muscle leading to the dilation of airways. However, their Japanese-language publication did not attract any interest in the West, nor did an ephedrine-containing preparation called Asthmatol marketed in Mukden, Manchuria, for the treatment of asthma. The veil on ephedrine suddenly lifted in 1924 with publication of work by K. K. Chen at the Peking Union Medical College.[7]

Chen grew up in China and attended the Tsing Hua College of Peking, where he learned English. He subsequently transferred to the University of Wisconsin, graduating in pharmacy, and spent two years in medical school before switching to a PhD program, where he studied biochemistry and physiology.

In 1923, Chen returned to China, joining the faculty of Peking Union Medical College, which had been established several years earlier with support from the Rockefeller Foundation. Chen had a major interest in identifying the active chemicals in herbal remedies; when a relative told him about the traditional properties of ma huang, he purchased some of the herb at a local shop and quickly purified ephedrine, only belatedly learning that Nagai had scooped him thirty years earlier. Nonetheless, Chen moved forward with an investigation of the pharmacology of ephedrine in animals and humans in collaboration with Carl Schmidt, an American pharmacologist on a several-year assignment in Peking. They rediscovered that ephedrine was a sympathomimetic drug. Chen and Schmidt did experiments with ephedrine on themselves and hospitalized patients, concluding that ephedrine demonstrated favorable results and displayed no problematic side effects. They appreciated that ephedrine, in comparison to adrenaline, had a longer duration of action and retained activity when given by mouth. We now know that ephedrine is effective orally because it is not metabolized by MAO. In 1924, their publication in a highly regarded and widely read American pharmacology journal brought wide attention to ephedrine in the West.[8]

In 1926, China exported about 220,000 pounds of ephedra to the United States, increasing exports to more than 1 million pounds by 1928. Drug companies around the world began marketing ephedrine. Reflecting the enormous volume of research on ephedrine that was done in just a few years, Chen and Schmidt wrote a massive 115-page review article in 1930 that primarily documented recent research with ephedrine. Ephedrine quickly became very popular in the treatment of asthma, hay fever, and colds.[9]

Medically unsupervised use of ephedrine (and its source, ephedra) for weight loss, fatigue, and the enhancement of athletic performance later raised many safety concerns.[10] The amount of ephedrine in ephedra is highly variable, unlike pharmaceutical-grade ephedrine, which is labeled with an accurate, specific dose. Ephedra-containing dietary supplements have been linked to serious adverse effects, including major mental disturbances and dangerous heart rhythm problems. Enormous adverse publicity regarding ephedra occurred in 2003 with the sudden death of Steve Bechler, a twenty-three-year-old pitcher in the Baltimore Orioles baseball organization.[11] Despite considerable lobbying from the dietary supplement industry, the FDA banned ephedra-containing food supplements in 2004. Ephedrine, however, remains available without a prescription.[12]

A Speedier Adrenaline: Amphetamine

Gordon Alles was working in an allergy laboratory in Los Angeles at about the time ephedrine became widely used for asthma.[13] With ephedrine in short supply and expensive, Alles decided to invent a novel drug for asthma and nasal congestion by synthesizing adrenaline analogs. He synthesized phenylisopropylamine, later given the generic name *amphetamine*. In animal experiments, amphetamine proved active in raising blood pressure after oral administration.

Alles moved forward quickly with human testing, serving as the first normal volunteer. A physician friend injected him with amphetamine under the skin; Alles's blood pressure rose, as predicted by the animal experiments. In addition, Alles noted that he felt enormously energetic and elated, and he had difficulty sleeping that night. Just two days later, a patient suffering from an asthma attack was given an oral dose of amphetamine. While the patient felt exhilarated, the asthma attack did not improve. Based on his discoveries, and after finding out that amphetamine had been synthesized decades earlier, Alles applied for a new-use patent for amphetamine.[14] Alles also synthesized the adrenaline analog methylenedioxyamphetamine (MDA), a chemical later misused as a psychedelic drug.[15]

While amphetamine was not very efficacious in asthma, human trials demonstrated that the drug made people feel good and gave them increased energy with less sleep. Physicians began using amphetamine to treat patients with excessive sleepiness and with depression. Amphetamine in combination with a sedative drug became a popular treatment for obesity; amphetamine curtailed appetite, while the sedative decreased overstimulation from amphetamine. Students began using amphetamine to stay awake while cramming for exams. Amphetamine unfortunately proved habit-forming and produced drug dependence, features of a potentially addictive drug. However, amphetamine remains a therapeutic option in patients with narcolepsy or attention deficit/hyperactivity disorder.[16]

Amphetamine's major effects are mediated indirectly by inducing the release of several neurotransmitters. Injections of the drug can cause severe elevations of blood pressure as a result of the dramatic release of noradrenaline from sympathetic nerve endings. In the brain, amphetamine causes its pleasurable effects by releasing neurotransmitters, especially dopamine.

Families of Adrenergic Agonists and Antagonists

Tyramine, clonidine, and amphetamine were initially one-of-a-kind sympathomimetic drugs whose unanticipated effects led to novel therapeutic indications or provided clues for later discoveries of fundamental biomedical importance. We now turn to pharmacological programs that exploit the favorable properties of a drug by designing a series of new analogs for a predetermined purpose. These strategies have been informed by the increasingly sophisticated understanding of the biology of adrenergic receptors, leading to the invention of drugs with improved therapeutic profiles or diminished adverse effects. Nonetheless, no matter how rational are the steps in both research and development, many of these drugs illustrate the considerable vigilance that is required to detect unanticipated adverse effects of any new drug.

More Selective Sympathomimetic Drugs for Asthma

Adrenaline is efficacious in asthma; however, its beneficial effects wear off quickly and adverse effects include changes in blood pressure and heart rate. After Ahlquist's demarcation between α and β adrenergic receptors, pharmacologists realized that adrenaline's beneficial effects in asthma were mediated by β receptors, whereas activation of α receptors contributed to elevated blood pressure via vasoconstriction. Even more refined pharmacological experiments several decades later suggested that selectively targeting β_2 receptors in bronchial smooth muscle would produce efficacious responses in asthma, while avoiding the activation of β_1 receptors in the heart would decrease cardiac adverse effects.[17]

In the 1930s, chemists empirically learned that adding bulky chemical groups to adrenaline's nitrogen atom produced new chemicals that dilated airways but had less capacity to constrict arteries. In the late 1930s, chemists at a drug company in Germany now known as Boehringer Ingelheim synthesized isoproterenol, an analog with an isopropyl group attached to adrenaline's nitrogen atom. Pharmacologists at the University of Vienna demonstrated that isoproterenol was much more potent than adrenaline in dilating airways in the lung and in increasing heart rate; even more remarkably, isoproterenol did not cause vasoconstriction. In modern terms, isoproterenol is a potent agonist at β_1 and β_2 receptors but has little capacity to activate α receptors.

Isoproterenol represented a pharmacological advance in the treatment of asthma: potently opening airways without causing sharp rises in blood pressure. Available in Germany during World War II, isoproterenol remained relatively unknown elsewhere until after the war. Administering aerosolized isoproterenol into the lungs apparently limited adverse cardiac effects. Containers that provided measured doses of isoproterenol under pressure made it easier for asthmatics to administer the drug into their lungs. Consequently, even without a sophisticated conceptual framework for adrenergic receptors, ingenious chemists and pharmacologists had invented a drug with desirable properties that seemed to be an important step forward in the treatment of asthma.

An aphorism attributed to William Osler in the early twentieth century, "The asthmatic pants into old age," made the point that asthma, while intermittently disabling, apparently only occasionally killed young patients. However, in the 1960s an epidemic of thousands of unexpected asthma deaths began in Western Europe, New Zealand, and Australia, but not in the United States or Canada. These deaths attracted the interest of epidemiologists—scientists concerned with understanding causes of disease in large populations. Their research suggested that the increased use of a highly concentrated form of isoproterenol—available in the countries with the epidemic but not in the United States or Canada—was responsible for the increased death rate. Some termed this association the "asthma paradox."

Many physicians, as well as pharmaceutical companies marketing isoproterenol, did not accept the conclusion that isoproterenol caused the epidemic. Alternative explanations for the increased death rate included the "delay hypothesis," which proposed that deteriorating asthmatics did not visit their physicians quickly enough because they felt better after using isoproterenol. Some physicians speculated that the severity of asthma was coincidentally increasing. However, others were concerned that isoproterenol in this concentrated form either worsened asthma or had inherent toxicity that led to fatal cardiac events in these patients. In any case, the death rate in asthmatics dropped after the highly concentrated isoproterenol formulation was removed from the market. While this outcome provides suggestive confirmatory evidence implicating isoproterenol, simultaneous educational efforts to improve treatment of asthma may have contributed to the drop in fatalities. This unfortunate episode emphasizes how challenging it is to recognize that a drug may simultaneously improve symptoms and increase death rates from a disease.

The next major advance with adrenaline analogs involved the synthesis of drugs that preferentially activated the β_2 receptors in the airways while tending to spare β_1 receptors in the heart. In the 1960s, terbutaline and albuterol (the latter called salbutamol in some countries) emerged as drugs more potent at β_2 receptors than at β_1 receptors.[18] Albuterol remains one of the most popular β_2 receptor agonist aerosols for asthma.[19]

In the classification of sympathomimetic drugs for asthma, adrenaline (equally potent at β_1 and β_2 receptors and active at α receptors) is a first-generation drug, isoproterenol (equally potent at β_1 and β_2 receptors but weak at α receptors) is a second-generation drug, and albuterol and related β_2-receptor-selective drugs are the third generation. In the late 1980s, salmeterol and formoterol were identified as fourth-generation β_2-receptor-selective agonists; these drugs have the advantage of acting for a long time due to prolonged residence near the β_2 receptors in the airways. While not recommended as sole treatment for asthma, these two drugs provide a baseline of favorable effects.

Quenching Adrenaline with α Adrenergic Receptor Antagonists

Just after World War II, Louis Goodman, the new chair of pharmacology at the University of Utah, collaborated with Mark Nickerson to work on dibenamine, a novel α receptor antagonist that blocked adrenaline-mediated vasoconstriction. Dibenamine is structurally related to the nitrogen mustards that Goodman and Alfred Gilman had worked on during World War II. Nickerson discovered that dibenamine irreversibly blocked vasoconstriction mediated by adrenaline. The effects of dibenamine lasted for days because it forms very stable chemical bonds with α adrenergic receptors, permanently preventing adrenaline from attaching to these receptors. Several days are required for cells to synthesize a normal complement of new α adrenergic receptors; if new receptors were not synthesized, the effect of this antagonist would not wear off. Phenoxybenzamine, more potent that dibenamine, also irreversibly inactivates α receptors. Physicians soon began using phenoxybenzamine to treat patients with pheochromocytomas prior to surgery, in order to ameliorate the hypertension caused by excessively high concentrations of adrenaline and noradrenaline that markedly stimulate α receptors in blood vessels.

In the 1970s, prazosin emerged as the first reversible antagonist of α_1 receptors that had almost no capacity to bind to α_2 receptors. The marked selectivity for α_1 receptors represented an important step forward, since blockade of α_2 receptors caused significant adverse effects with some earlier drugs.[20] Prazosin and a number of similar analogs gained wide acceptance as drugs that lowered blood pressure in patients with hypertension. Several decades after prazosin's approval in the United States, a large clinical trial investigating a closely related drug raised concern that it might not be as effective in preventing heart failure as drugs with different mechanisms of action. Subsequently, enthusiasm for prazosin and related drugs as the sole treatment for hypertension waned considerably.[21]

However, prazosin and related α_1 adrenergic receptor selective antagonists have another action that retains considerable benefit: they are efficacious in improving unpleasant urinary symptoms in men with benign prostate hyperplasia. In this condition, the enlarged prostate compresses the urethra and impedes the flow of urine. Administering α_1 receptor antagonist drugs decreases noradrenaline-induced contraction of muscles in and near the prostate, leading to improved urine flow.[22]

Quenching Adrenaline with β Adrenergic Receptor Antagonists

The treatment of angina with propranolol has benefited millions of patients. Since the introduction of that drug, more than fifteen additional β receptor antagonists have been marketed around the world; some of these drugs constitute important upgrades. The earliest major improvement involved the development of β receptor antagonists that were more potent at β_1 than at β_2 receptors—in other words, drugs with greater effects on the heart (β_1 receptors) than the lungs (β_2 receptors), contributing modestly greater safety in some patients with difficulty breathing. Later-generation β receptor antagonists have supplemental pharmacological actions including blocking α_1 receptors or dilating blood vessels through other mechanisms. Because propranolol lasts only a short time in the body, patients with angina originally had to take the medication three or more times daily to get around-the-clock relief from angina, but newer β receptor antagonists can be used just once daily. This advance required either the invention of novel β receptor antagonists that inherently last a long time in the body or the development of new formulations of older short-acting drugs such as propranolol that are absorbed more slowly from the intestines. On the other hand, some very sick patients can develop life-threatening complications immediately after a single dose of a β receptor antagonist. For these patients, an intravenous infusion of the novel β receptor antagonist esmolol may be the safest option since it is eliminated from the body in mere minutes.

Drugs with improved pharmacological properties are not necessarily better drugs. The story of the β receptor antagonist practolol illustrates the challenges in recognizing uncommon but very serious adverse effects, especially after a drug is marketed. ICI discovered that practolol was selective for β_1 receptors compared to β_2 receptors. This represented a potentially major therapeutic advance: a β receptor antagonist for angina that might have fewer adverse effects in the lungs. In 1970, ICI marketed practolol in the United Kingdom and partnered with an American company to secure approval in the United States. The FDA had reservations because practolol caused cancer in laboratory animals, and the drug never reached the U.S. market.

Physicians in the United Kingdom began reporting that patients prescribed practolol had very severe and sometimes bizarre reactions involving the skin, eyes, and membranes that surround major organs. Thousands of patients had major adverse effects that included blindness, and dozens died. ICI withdrew practolol five years after its launch. The damage from practolol has been viewed as a disaster in the same league as the earlier devastating birth defects caused by thalidomide.[23]

Initially, many believed that the practolol tragedy occurred because of rare and unpredictable outcomes not noted in animal testing or in clinical trials. However, in retrospect, patients in the trials reported complaints involving their eyes and skin that went unrecognized as adverse drug effects. Valuable lessons learned from this include the importance of rigorous analysis of clinical trial data and the need for careful follow-up of new drugs after they reach the market. Unfortunately, even thirty-five years after the bitter experience with practolol, critically important surveillance for adverse effects after a drug is marketed remains less than optimal around the world.

Expanded Clinical Uses of β Receptor Antagonists

Clinical uses for β receptor antagonists have increased enormously since their introduction for angina in the 1960s. In clinical trials involving patients with angina, investigators noted that propranolol lowered blood pressure in patients with co-existing hypertension. Nonetheless, acceptance of the idea that a drug for angina could be an antihypertensive drug encountered widespread resistance because this action was unexpected. Investment in almost a decade's worth of clinical research and drug trials finally overcame skepticism about prescribing β receptor antagonists for hypertension.[24] Eventually β receptor antagonists became frontline drugs in hypertension. Curiously, despite literally thousands of scientific publications involving β receptor antagonists, their precise antihypertensive mechanisms remain uncertain.

Another major expansion in the use of β receptor antagonists has been for patients after heart attacks. Heart attack survivors who are discharged from the hospital with a prescription for a β receptor antagonist live longer. Unfortunately, even a decade after a powerful clinical trial demonstrated this drug's remarkable efficacy in saving lives, far too few eligible heart attack patients in the United States were receiving a prescription for a β receptor antagonist. Translating good outcomes in clinical trials into routine practice is very difficult even when the therapy is effective, safe, and inexpensive. The failure to widely prescribe β receptor antagonists in heart attack survivors raised concerns nationally. In the mid-1980s, organizations measuring the quality of medical care and accrediting hospitals began to focus attention on the proportion of heart attack survivors who received prescriptions for β receptor antagonists. Over the course of several decades, the combined efforts of these organizations, medical insurance companies, and widespread educational efforts—coupled with turnover in the physician workforce—finally led to the extensive use of β receptor antagonists in heart attack survivors. While many people look forward to new therapies based on revolutions in genomics and stem cell biology, we could immediately improve the treatment of disease if well-established and inexpensive therapies were more widely and quickly offered to patients.[25]

The demonstration that β receptor antagonists prolong the lives of patients with congestive heart failure epitomizes a major yet counterintuitive clinical advance. The ravages of heart attacks, severe hypertension, and other diseases can lead to heart failure, which is a syndrome characterized by shortness of breath and decreased life expectancy. With heart failure, sympathetic nervous system activity typically rises, stimulating a weakened heart to pump harder. Physicians quickly learned that a strong dose of a β receptor antagonist in patients with severe heart failure could acutely worsen symptoms by diminishing the stimulatory effects of catecholamines on the heart. Indeed, in life-threatening episodes of heart failure, physicians often temporarily prescribe adrenaline-related agonists to stimulate the diseased heart to pump at its maximal capacity. However, over the course of weeks or months, these agonists typically stop working and likely shorten life expectancy, as flogging a weakened heart is ultimately counterproductive. Nonetheless, the counterintuitive discovery that the cautious use of β receptor antagonists can prolong life in some patients with heart failure represents a striking and imaginative therapeutic advance. These drugs are now standard treatment for heart failure in patients with diminished cardiac pumping capacity; initiating treatment with very low doses has proved relatively safe. Long-term use of β receptor antagonists may reverse some of the consequences of overstimulation of β receptors that occur in heart failure due to chronic activation of the sympathetic nervous system.

The efficacy of β receptor antagonists in heart failure demonstrates that the long-term effects of a drug may be very different from its short-term actions. Physiological adaptations to drugs over time may have many consequences—desirable or undesirable—for patients. Lessons from heart failure have stimulated early experimental interest in the possibility that β receptor antagonists might even be useful in the treatment of asthma.

Over the years, through accident or design, β receptor antagonists have found unanticipated uses outside the cardiovascular system. Glaucoma impairs vision by damaging the optic nerve, often in association with high pressure in the fluids in the eyes. Several years after marketing of propranolol for angina, physicians noted that propranolol lowered pressure in the eyes. In the late 1970s, timolol eye drops became the first β receptor antagonist approved for the treatment of glaucoma. After timolol became widely used, however, physicians learned from hard experience that substantial amounts of timolol can be absorbed from the eyes into the bloodstream, leading to blockade of β_2 receptors in the lungs, with the risk of precipitating attacks of asthma in susceptible patients.

Some β receptor antagonists are useful in preventing migraine attacks, decreasing shaking of the hands in patients with familial tremor, decreasing symptoms of an overactive thyroid gland, decreasing unpleasant symptoms in alcoholics undergoing acute withdrawal, and preventing gastrointestinal bleeding from enlarged veins (varices) that develop in patients with cirrhosis of the liver.[26] The discovery of the efficacy of β receptor antagonists in these and other clinical settings typically has involved some combination of observant physicians and good pharmacological reasoning.

Another use of β receptor antagonists is to ameliorate performance anxiety, especially in musicians. The stress of a major performance, for example, can make delicate fingering a challenge for a violinist, because an adrenaline surge can increase hand tremors by activating β receptors in skeletal muscles. To combat this normal response and perform with greater confidence, some performers take a β receptor antagonist before going onstage. Blair Tindall wrote that these drugs are an open secret among musicians.[27] A player in a major orchestra suggested that musicians did not typically disclose using propranolol, much the way that men do not volunteer the need for a drug in cases of erectile dysfunction. Several placebo-controlled trials with musicians have confirmed the impression that β receptor antagonists improve musical performance. However, practices that may be acceptable for musicians are not tolerated in athletes. In major competitions such as the Olympics, archers and other shooters are routinely screened for illicit use of β receptor antagonists, which are viewed as performance-enhancing drugs in these sports. By attenuating slight natural tremors, β receptor antagonists can improve shooting accuracy. This impression has been confirmed in placebo-controlled trials with amateur marksmen. A pistol shooter lost his two medals and was expelled from the Olympics in 2008 after testing positive for propranolol. On the other hand, β receptor antagonists might be a natural choice for elite military snipers planning a very difficult shot.

[TEN]

Adrenaline Junkies

Adrenalin Mother,
with your dress of comets
and shoes of swift bird wings
and shadow of jumping fish,
thank you for touching,
understanding and loving my life.
Without you, I am dead.

RICHARD BRAUTIGAN

At just the right dose, adrenaline connotes excitement, vigor, and thrills. On the other hand, overflowing adrenaline heralds fear, anger, and even death. Similarly, many of the ways adrenaline has been portrayed over the years in popular media involve its capacity to increase strength and to boost the flavor and intensity of emotional responses; other allusions draw upon its ability to trigger heart attacks.

Allusions to Adrenaline

The most fashionable facets of adrenaline have changed considerably over the last hundred years. The first story in the *New York Times* involving adrenaline appeared on the front page on Sunday, January 18, 1903. The article described an advance that would "astonish the medical profession throughout the world": George Crile had revived a dog with an injection of adrenaline fifteen minutes after its heart had stopped. In the following decades, newspaper articles involving adrenaline mainly addressed biomedical advances, particularly in resuscitating the dead. However, in the 1930s and 1940s, the word *adrenaline* came to connote energy for individuals, groups, even entire organizations, and "a shot of adrenaline" became a cliché for infusing energy. The use of *adrenaline* in nonmedical contexts took off in the 1960s in stories about sports and thrill seeking. The increasing appearance of *adrenaline* in a newspaper is illustrated here; after 1940, medical citations represent only a very small usage in these many articles.

While newspaper coverage through the 1920s highlighted adrenaline's therapeutic use in reviving people, often with naive expectations, movies and books enormously exaggerated adrenaline's capacity to raise the dead. For example, the 1923 movie *Legally Dead* tells a complicated story of a reporter hanged for a murder that he did not commit; a physician brought him back to life with adrenaline to overcome the injustice. In the 1929 novel *Murder by the Clock*, a young man is strangled to death; hours later a physician revives him with adrenaline. More recently, in the 1994 hit movie *Pulp Fiction* the gangster Vincent (John Travolta) is terrified when his boss's wife, Mia (Uma Thurman), apparently dies after taking an illicit drug. The well-prepared local drug dealer gives Vincent an enormous syringe filled with adrenaline, instructing him to inject directly into Mia's heart through her breastbone. Vincent dutifully stabs her chest; moments later, without further intervention, Mia is back to normal.

These depictions of adrenaline's capacity to resurrect the dead involve wildly implausible medical scenarios. In reality, adrenaline, in conjunction with mechanical cardiopulmonary resuscitation and electric shocks, is used to restart hearts that have stopped beating. However, cardiac resuscitation should start within minutes to prevent irreversible brain damage; adrenaline alone does not generally restore the circulation without additional interventions.

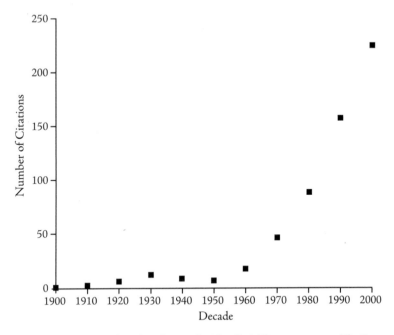

Figure 10-1. Citations for adrenaline in the *New York Times* newspaper. The Pro-Quest Historical Newspapers database for the *New York Times* from 1851 to 2007 was searched using the key words *adrenaline* and *adrenalin.* The average number of citations per year in each decade starting from 1900 is plotted on the graph. For the final decade, the number reported is averaged through the end of 2007. Up until the 1960s, mention of adrenaline appeared in this newspaper only a few times each year; since that time, the number of citations has grown considerably without any indication of leveling off.

The miraculous capacity of adrenaline to restore life was quickly extrapolated to animating the inanimate. In 1923, Will Rogers wondered what effect adrenaline would have on dead campaign issues. The *Washington Post* noted in 1928 that in the film industry, microphones were like injections of adrenaline into the hearts of dead people: injecting sound revived terrible, unreleased silent movies, allowing them to come off the shelves with enough dialogue to qualify as "talking pictures." The *New York Times* reported in 1947 that the appearance of the future Hall of Fame pitcher Dizzy Dean in a one-game comeback with the St. Louis Browns baseball team on the last day of the season acted "as a shot of adrenalin to a dying box office," bringing in a total of 15,916 fans—"approximately 15,000 more than would normally have been at hand." After almost a hundred years, the metaphor of adrenaline as a force that could bring just about anything back to life has been beaten to death many times over.

The capacity of "a shot of adrenaline" to increase energy and vigor has been exploited in writing about sports, politics, and other entertainments. The *Chicago Daily Tribune* ran a headline in 1923 on the prospects of Yale University's baseball team: "Yale's Nine Has Material but Needs Shot of Adrenalin." In the same year, the *Washington Post* wondered if cannibals had the right idea about gaining strength by eating their enemies' adrenal glands, the glands of courage. A columnist in the *Washington Post* noted in 1926 that Dr. John Wilce, the coach of the Ohio State University football team, had stated, "It is the adrenal gland which does the trick when a football player gets keyed up." Wilce, later a professor of preventive medicine, cautioned that he dared not key up his team more than twice a season, and the columnist speculated that coaches could be spared the need for oratory by simply shooting adrenaline into their players, with the dose proportionate to the caliber of the opposition. On June 16, 1940, the *Post* reported about interventionists who wanted a declaration of war from the United States to join the Allies against Nazi Germany; one of their justifications included giving "the Allies a psychological shot of adrenaline which is indispensable to their survival for another few months." As it happened, the adrenaline was withheld and France signed an armistice with Germany about a week later.

Syndicated columnist Bob Considine was an early adopter of metaphorical adrenaline in sports stories. In 1941, many speculated about the financial value of the New York Yankees. Considine wrote that while the price might have gone down as the club dropped into fifth place, Joe DiMaggio's record hitting streak of fifty-six consecutive games "shot a quart of adrenalin in the club's veins . . . [changing] the outlook." Two years later, Considine wrote that American track and field had had many successes but owed much to a "handful of foreigners who have shot adrenaline into our track and field world at many periods when its pulse was low." Arthur Daley, sports reporter and later Pulitzer Prize–winning sports columnist for the *New York Times*, was also an early adopter of *adrenaline* in a sports context. In 1944, he did a column on Gus Sonnenberg's contributions to professional wrestling, crediting him with naming fancy holds such as the Airplane Spin and the Bear Hug. Daley explained that despite the rigged matches, Sonnenberg's flying tackle "gave a shot of adrenaline to the wrestling box office." In 1945, he wrote a column about how the skill level in college sports, especially football, had diminished during World War II; he had great hope for a revival, as he thought that "V-E Day is going to act like a shot of adrenaline on a very sick patient." In 1965, *Time* magazine reported that Jack Nicklaus drove a ball more than 350 yards in winning the Masters Golf Tournament, and quoted him as saying, "My adrenalin is running strong, I'm all pumped up inside."[1]

Arguably, the most remarkable, transformative actions of adrenaline favoring energy and strength are those involving the Incredible Hulk, a well-known cartoon, TV, and movie character.

The process by which Banner usually transforms into the Hulk has a chemical catalyst, adrenalin (aka epinephrine). As in normal human beings, Banner's adrenal medulla secretes huge amounts of adrenalin in times of fear, rage or stress, which hormonally stimulates his heart rate, raises blood-sugar levels and inhibits sensations of fatigue. Whereas the secretion heightens natural physical abilities in normal human beings, in Banner's case it triggers the complex chemical extra-physical process that transforms him into the Hulk. The total transformation takes anywhere from seconds to as long as 5 minutes, depending on the initial adrenaline surge, which is determined by the original external stimulus. Soon after the transformation, Hulk's adrenalin levels will return to more normal, reduced levels. At various times in his life, Banner has been able to mentally control the change, producing or inhibiting it in either direction. At many times in the past, Banner would revert to the Hulk, regardless of the time of day or night, whenever he underwent enough excitation to induce a sufficient surge of adrenalin to trigger the transformation. Similarly, often if the Hulk relaxed to a great enough degree, his adrenalin levels would decline and he would revert to human form; however, there are many known instances in which the Hulk was very relaxed or even slept without reverting to human form. . . . The Hulk's adrenalin levels counteract fatigue poisons; while fighting others in an enraged state, he can maintain his peak output for hours on end and still continue to grow even stronger as his anger escalates. He has swum across both Atlantic and Pacific oceans, though he did become extremely exhausted in the process, presumably from lack of anger.[2]

The capacity of adrenaline to energize is depicted at a much more modest level in the movie *Mission: Impossible III*. A secret agent (played by Tom Cruise) injects a large dose of adrenaline into his protégé (played by Keri Russell) after she has been beaten into a stupor by the accomplices of an evil arms dealer. Almost immediately she jumps up, fully energized, to help her rescuers in an intense gun fight before escaping.

More Power to You

Widespread news reports suggest that even ordinary people can transiently have superhuman strength in response to great danger, with stories of citizens lifting cars off injured children and outrunning or defeating wild beasts including polar bears. Many of these feats have been attributed to "adrenaline rushes." In full disclosure, I experienced something along those lines myself. As a plodding but enthusiastic fortyish cyclist with many miles in my legs, I was riding alone one day along a nearly deserted rural road in California at my cruising speed of 19 mph when a car whizzed by, passing within inches. Foolishly, I tapped my helmet, expressing an unnecessary opinion about the driver. He caught the gesture in his rearview mirror and immediately pulled off the road about fifty yards ahead. A stocky man leaped out of the car and started toward me on foot; he called out once, "You're going down." Somewhat alarmed, I balanced precariously on my bicycle, feet clipped into the pedals. After he had walked about fifty feet from his car toward me, I abruptly sped off in the opposite direction, hoping to reach a small refreshment stand about a mile or two away before he could return to his car and run me down. Glancing at the handlebars, I saw that the bike's speedometer read 30 mph and my heart rate monitor flickered around 190 beats per minute. I had never before cracked 25 mph except going downhill and had thought that my maximum heart rate would not exceed 180. At this pace, I expected to be exhausted within seconds, but I dared not ease up. Nonetheless, I felt no shortness of breath; this puzzling performance provided a welcome distraction from my immediate concerns. Within minutes I reached the stand, where people in the parking lot provided a measure of safety. Fortunately, my adversary had not bothered to pursue me. Twenty years later, I remain astonished by this unprecedented speed and output of energy, achieved so readily, that far surpassed my capacities before or since on a bicycle.

Along similar lines, the *Boston Globe* reported on July 9, 2012, about the response of first-time kayaker Walter Szulc Jr. after he realized that onlookers were frantically signaling that he was being followed by a very large shark just off a Cape Cod beach: "The chase lasted maybe 90 seconds, Szulc and witnesses estimated, and he said he was running on adrenaline and reactions. 'All I could sense is that I should paddle,' he said. 'I just knew that I didn't want to end up in the water. And paddling-wise, I turned into a professional kayaker all of a sudden. I paddled like there was no tomorrow, like my life depended on it, and it's quite possible that was the case.'"

Given the many informal endorsements that claim emotional responses increase brute force, what is known from a scientific viewpoint about adrenaline's capacity to increase physical performance? Considerable research in sports medicine laboratories has evaluated the role of adrenaline and sympathetic nervous system activity in intense running and cycling exercises. Concentrations of adrenaline and noradrenaline increase manyfold in the blood during these activities. Physically conditioned athletes secrete more adrenaline than untrained people; this augmented capacity has been called the "sports adrenal medulla." Limited evidence suggests that in trained athletes, adrenaline secretion in anticipation of exercise modestly augments muscle force at the onset of exercise. Moreover, β adrenergic antagonist drugs, which oppose the actions of adrenaline, impair maximal exercise capacity, especially by limiting cardiovascular and metabolic responses. Coupled with much additional evidence, this suggests compellingly that catecholamines make important contributions to an individual's maximum physical capacity. However, the possibility that a terrifying fight-or-flight stimulus would charge an unfit person with the capacity to lift an enormous weight off an injured child or run hundreds of yards at top speed to escape a mugger is untested in laboratory settings. Moreover, setting up surrogate emotional stimuli of that intensity in a controlled experiment would be very difficult to design and carry out ethically.

Adrenaline does have immediate effects that can increase muscle strength. Activation of α adrenergic receptors on the endings of nerves that cause muscle contraction enhances the release of the neurotransmitter acetylcholine, which may potentiate the resulting contractions. The capacity of muscle to contract depends on ATP, the chemical that provides energy to shorten muscle fibers. However, a leg muscle contains enough stored ATP to get a runner only partway down the track. To run longer distances, exercising muscle must burn glucose and other nutrients to regenerate ATP. As described in Chapter 7, adrenaline activates PKA, which stimulates the enzyme glycogen phosphorylase to release glucose from stored glycogen; this glucose contributes fuel needed to replenish stores of ATP. In addition, adrenaline has multiple effects on the body that improve the function of exercising muscles. Some of these actions include liberating glucose from the liver and fatty acids from adipose tissue (to provide additional fuel) and increasing the output of blood from the heart and dilating blood vessels going to muscles (which together enhance the delivery of nutrients to exercising muscle).[3]

Activation of β_2 receptors in skeletal muscle over time can increase muscle size and strength. PKA phosphorylates a single amino acid in a protein called cAMP response element binding protein (CREB). Through interactions with DNA sequences found in some genes, phosphorylated CREB leads to increased synthesis of proteins encoded by these genes; the buildup of these proteins increases muscle size.

Athletes have abused β_2 receptor agonists in the hope of improving muscle strength. Oral doses of these drugs are prohibited in the Olympics and other major events. Competitors with asthma are permitted to treat their disease with approved β_2 receptor agonists, but only if the medication is taken by inhalation in therapeutic doses. Clenbuterol is a selective β_2 receptor agonist that builds muscle and increases the growth rate of cattle; however, the drug is banned for use in cattle in many countries. A number of world-class athletes have taken clenbuterol in an effort to increase strength but have been caught and suspended after positive tests for this drug. Alberto Contador, a three-time winner of the Tour de France, tested positive for the drug in the 2010 Tour. He claimed that the drug got into his body inadvertently, as a consequence of unwittingly eating meat from an animal treated with the drug. Nonetheless, he was ultimately stripped of the 2010 title.[4] As noted in Chapter 9, competitors in shooting sports have exploited β receptor antagonists for unfair advantage. The abuse, in different sports, of drugs that either mimic or oppose the actions of adrenaline illustrates the high level of pharmacological sophistication and imagination of athletes who cheat. Alas, none of this is especially new. *Time* magazine reported in 1946 that at Loyola University's first turtle derby, a twenty-five-year-old terrapin named Arkansas Express ran away with the championship. He had been purchased at a fish market and trained for a month by premedical students. However, the report noted that there were no rules governing the competition and that one of the other contestants had dropped dead from an overdose of adrenaline.

Frightened to Death: Emotions and the Heart

The capacity of emotions to trigger sudden death and heart attacks is an established popular belief as well as an active area of medical investigation. *Trigger* in this context refers to an event that leads to a heart attack or sudden death within hours or days. Triggers are distinct from cardiac risk factors, such as smoking or elevated cholesterol, that increase the likelihood of a heart attack over the course of years. There is evidence that chronic emotional stress is a cardiac risk factor; however, our focus will be on the more dramatic triggers. What is the basis for the belief that emotional events can trigger heart attacks and sudden death? If there is a connection, how might adrenaline and the sympathetic nervous system potentially cause these devastating events?

The belief that intensive emotional experiences can trigger sudden death arose long before the phenomenon received attention from medical investigators. Acts 5:1–11 describes a forceful story involving sudden deaths: "Ananias, with Sapphira his wife, sold a possession, and kept back part of the price. . . . But Peter said, Ananias, why hath Satan filled thine heart to lie. . . . Thou hast not lied unto men, but unto God. Ananias hearing these words fell down, and . . . [died]. . . . Three hours [later] . . . his wife . . . came in. . . . Peter said unto her, How is it that ye have agreed together to tempt the Spirit of the Lord? Them which have buried thy husband . . . shall carry thee out. Then fell she down straightway at his feet, and [died]."

While not representing the strongest evidence, medical case reports support the concept that emotions can trigger sudden death. In the late eighteenth century, William Heberden, in the earliest description of angina, observed that angina occurred not only with exercise but also with "disturbance of the mind"; he noted that patients with angina often died suddenly. John Hunter, Heberden's famous surgical contemporary, developed exercise-induced angina. He also had angina in response to mental irritation, a major problem for a person who was intolerant of relatively minor upsets. He noted with foreboding, "My life is at the mercy of any scoundrel who chooses to put me in a passion." Hunter died suddenly at a contentious meeting with colleagues at St. George's Hospital in London. That morning, Hunter had made a prediction to a friend about the upcoming meeting: "Some unpleasant dispute might occur . . . [and] it would certainly prove fatal." His autopsy revealed advanced atherosclerosis in his coronary arteries. Through the nineteenth century in Europe, there was a growing sense that heart disease was increasing, possibly associated with greater emotional challenges in more crowded conditions. In 1910, the Russian physicians V. P. Obraztsov and N. D. Strazhesko presented their seminal work associating symptoms of heart attacks with coronary thrombosis (clots in the coronary arteries); they noted that emotionally charged events precipitated heart attacks.

In 1942, Walter Cannon at Harvard University wrote an influential paper about sudden death based on reports by anthropologists claiming that spells and other forms of "black magic" or "voodoo" led to death in "primitive" people from diverse cultures in South America, Africa, Australia, the Pacific Islands, and Haiti. Cannon identified numerous case reports of fear-induced death in native people of South America dating to the sixteenth century. He found a common feature in these cases: the belief that a powerful person could cause the death of defenseless victims. Cannon concluded that voodoo death may be real, precipitated by considerable emotional stress. He speculated that the autonomic nervous system mediated sudden death in the victims.

Cannon's laboratory research program had elucidated the major roles of adrenaline and the sympathetic nervous system in maintaining physiological balance and in responding to environmental stresses that represented emergencies. He believed that stress reactions involving adrenaline and the sympathetic nervous system provided a survival advantage by integrating physiological responses to threats such as dealing with beasts of prey. Hostile threats required intense physical responses, particularly holding one's ground or fleeing. While these concepts were somewhat controversial during Cannon's life, this framework is now widely accepted. These mechanisms presumably evolved over tens of thousands of years, enabling our ancestors to survive tough conditions. However, fight-or-flight responses, often coupled with rage, fear, and other emotions, may be counterproductive and maladaptive in our more modern society, where there are many frequent threats that carry symbolic significance but do not elicit physical exertion in response. Cannon believed that intense or prolonged activation of the sympathetic nervous system and adrenal medulla could have deleterious consequences for health.

In 1966, almost twenty-five years after Cannon's article on voodoo, a detailed medical report described the death in Baltimore City Hospital of a previously healthy twenty-two-year-old woman. Her chief complaints were shortness of breath and chest pain for a month. After two weeks in the hospital she finally revealed an unexpected concern. A physician described the situation:

> She had been born on Friday the 13th in the Okefenokee Swamp, aided by a midwife who delivered 3 children that day. The midwife told the mothers that the 3 children were hexed and that the first would die before her 16th birthday, the second before her 21st birthday, and the third (the patient) before her 23rd birthday. The first girl died in an automobile accident the day before her 16th birthday. The second girl went out to celebrate her 21st birthday and was killed by stray bullet. The patient firmly believed she was doomed and was manifestly terrified. . . . [O]ne day before her 23rd birthday, the patient died, following an episode of hyperventilation [excessive breathing], severe apprehension and profuse sweating.[5]

The autopsy demonstrated that the patient suffered from pulmonary hypertension, a disease associated with very high blood pressure in the artery carrying blood from the heart to the lungs. The physician discussing the case pointed to the hex as a factor in the patient's death, something the autopsy could not exclude. He speculated about the possible role of the autonomic nervous system along the lines suggested by Cannon decades earlier.

Both Cannon's review and the Johns Hopkins conference described reports of individual cases.[6] While dramatic and often seemingly persuasive, medicine is replete with anecdotal reports that appeared compelling but later proved misleading or wrong. George Engel took a further step with a systematic review of multiple cases of sudden death associated with emotions to discern potentially important common features. Engel had major interests in the effects of psychological and social factors in disease. In 1971, he analyzed 170 press reports of death that occurred after a dramatic life event. Most of the men involved were between forty-five and fifty-five years of age, and most of the women between seventy and seventy-five. He grouped the life events into several categories: losses (death of someone close, mourning, or anniversary of loss), personal danger, and happy events (reunions, triumphs). The frequency of deaths in these categories was 59 percent for loss, 34 percent for danger, and 6 percent for happiness. Men seemed to die most from danger, while for women loss was a more prominent category. An exception was loss of self-esteem, which included nine men but no women. As an example of the connection between emotions and sudden death, Engel described the case of the twenty-seven-year-old army captain who had commanded the ceremonial troops at the funeral of President Kennedy; the soldier subsequently died of a cardiac irregularity and heart failure ten days after the president, suggestive of the capacity of a challenging event to either damage the heart or cause life-threatening arrhythmias.

Analyzing data from case reports about sudden death has enormous heuristic value, motivating further research using more rigorous methods in epidemiology to test the hypothesis that emotions trigger major cardiovascular crises. Epidemiology is the scientific discipline broadly concerned with asking questions about health in large populations using powerful statistical techniques to evaluate factors that cause disease; in general, epidemiological studies are more compelling than accumulated case reports. Some of these methods have been used to test the hypothesis that emotions trigger cardiovascular events, including sudden deaths and heart attacks. In 1975, a stringent study from England examined conditions surrounding the sudden deaths of one hundred patients with coronary artery disease compared to a control group. Acute psychological stress occurred much more frequently in the people who died suddenly, confirming that emotional reactions contribute to death in patients with heart disease. In 1990, a large NIH-supported study in the United States demonstrated that almost 20 percent of patients with a heart attack reported a trigger of a mentally stressful event prior to their attack.

An alternative approach to this question has asked whether public events that simultaneously stress thousands of people are associated with excessive cardiovascular events in the exposed population. Terrorism, earthquakes, stock market volatility, and major sporting events have been assessed as potential triggers. Hospital admissions for heart attacks increased near Tel Aviv, Israel, during Iraqi missile attacks on civilians in the first week of the Gulf War in 1991. However, hospital admissions for acute cardiac events in New York City did not increase following the terrorist attacks on the World Trade Center on September 11, 2001. On the other hand, there was a doubling of life-threatening arrhythmias in patients with serious heart disease at that time.

In 1981, a 6.7-magnitude earthquake shook Athens, Greece, followed by multiple aftershocks for several days. Cardiac deaths increased during the period of the aftershocks. In 1994, an earthquake of similar magnitude struck near Los Angeles, California; investigators identified twenty-four cases of sudden death on the day of the earthquake, markedly more than the usual five cases a day. In addition, hospital admissions for heart attacks increased for several days following the earthquake.

In 2011, Chinese investigators reported on the relationship between the daily change in the Shanghai Stock Exchange Composite Index and deaths from coronary heart disease between 2006 and 2008. During that time, the Chinese stock market had periods of remarkable growth (up from 1,164 to 6,124) as well as collapse (back down to 1,708). They found that both rising and falling values of the index were associated with increased deaths; the fewest deaths occurred when the index was relatively stable. Each 100-point change in the index was associated with about a 5 percent increase in cardiac deaths.

In 1998, England lost to Argentina in a penalty shoot-out in a round-of-sixteen World Cup soccer match. On game day and for the next two days, hospital admissions in England for heart attacks increased. In the 2006 World Cup, broadly similar findings were reported when the host country, Germany, played its matches.

While observational studies in populations have limitations, overall they support the hypothesis that severe stress is associated with heart attacks and sudden death. However, the fact than one event precedes another does not imply linkage by a cause-and-effect relationship; the early morning crowing of a rooster does not cause the sun to rise.[7] Consequently, the results of carefully designed experiments testing whether stress can induce cardiac ischemia (inadequate blood flow to the heart muscle) are especially meaningful.

Experimental stresses induced in laboratories, such as having subjects with coronary artery disease do mental arithmetic under pressure or simulate public speaking, can provoke cardiac ischemia comparable to changes induced by vigorous physical exertion. Moreover, monitoring patients with coronary artery disease for ischemia during ordinary daily activities has demonstrated similar ischemic changes induced by tension, frustration, or anger. There is some evidence that people who survive a heart attack triggered by emotions have more intense blood pressure responses to experimental psychological stress compared to people who had an unprovoked heart attack. The results of many studies provide strong support that even relatively mild emotional stress can cause cardiac ischemia in many patients with coronary artery disease.

There is now considerable evidence that the brain can activate physiological responses that may damage the heart. Walter Hess and later investigators identified brain centers that respond to emotionally stirring events by turning on activity in the sympathetic nervous system and enhancing adrenaline secretion. These responses evolved so that our ancestors could most effectively deal with serious environmental threats. These pathways increase the work of the heart, leading to enhanced oxygen requirements in heart muscle. In healthy people, dilation of coronary arteries brings in oxygen-rich blood to amply meet these needs. However, in people with atherosclerotic coronary arteries, catecholamines can cause coronary artery vasoconstriction (rather than the typical vasodilation), which further diminishes blood flow to the heart muscle. The combination of diminished coronary blood flow and increased workload of the heart can precipitate ischemia of the heart muscle, making the heart more prone to cardiac rhythm disturbances that sometimes terminate in cardiac arrest.[8]

Fundamental research on the biology underlying heart attacks has revealed novel connections between emotions and adverse cardiovascular responses. The immediate precipitating event of most heart attacks is the sudden rupture of an atherosclerotic plaque lining a coronary artery. Exposed materials in the plaque cause platelets and clotting factors in the blood to form a large clot that obstructs blood flow. The section of heart muscle that ordinarily receives blood from that coronary artery soon dies if the obstruction is not reversed very quickly. Adrenaline and the sympathetic nervous system may increase the risk of the rupture of a plaque. The mechanism for this may be related to catecholamine-induced constriction of coronary arteries coupled with a rise in blood pressure; these changes increase turbulent blood flow which stresses the walls of the coronary arteries, increasing the likelihood of a plaque rupture.[9]

Catecholamines have additional effects that may increase the risk of a clot in the coronary arteries. For example, adrenaline activates α_2 receptors in platelets, which primes them to plug arteries.[10] In addition, catecholamines increase the concentration of several clotting factors, enhancing the capacity of blood to coagulate. While stemming blood flow is desirable during fight-or-flight responses that may be associated with bleeding from wounds or bites, these mechanisms are counterproductive in an emotional response in a person with atherosclerotic coronary arteries. When there is a ruptured atherosclerotic plaque, activated platelets and enhanced blood clotting mechanisms worsen the risk that a clot will completely obstruct a coronary artery.

In summary, considerable evidence supports the widely held impression that people can drop dead from events that stimulate intense emotional responses. This notion has been confirmed not only with large numbers of case reports but also by sophisticated epidemiological studies. Clinical experiments have substantiated this hypothesis. Moreover, fundamental research has revealed biological effects of adrenaline and noradrenaline that provide mechanistic understanding of their role in contributing to the risk of heart attacks and sudden death. Nonetheless, all is not doom and gloom with effects of emotions. The next section takes up the role of adrenaline in how we feel, especially in association with excitement.

Emotions Flavored by the Spice of Adrenaline

In commercial contexts, the word *adrenaline* is often employed to suggest a desirable state of excitement to enhance the attractiveness of merchandise, activities, or locations. The word has appeared in trademarked names for everything from fitness centers to sunglasses.[11]

Aldous Huxley's novel *Brave New World*, published in 1932 and set about five hundred years in the future, describes the utility of adrenaline in a changed society:

> "There's a great deal in it," the Controller replied. "Men and women must have their adrenals stimulated from time to time."
>
> "What?" questioned the Savage, uncomprehending.
>
> "It's one of the conditions of perfect health. That's why we've made the V.P.S. treatments compulsory."
>
> "V.P.S.?"
>
> "Violent Passion Surrogate. Regularly once a month. We flood the whole system with adrenin [adrenaline]. It's the complete physiological equivalent of fear and rage. All the tonic effects of murdering Desdemona and being murdered by Othello, without any of the inconveniences."
>
> "But I like the inconveniences."
>
> "We don't," said the Controller. "We prefer to do things comfortably."
>
> "But I don't want comfort. I want God, I want poetry, I want real danger, I want freedom, I want goodness. I want sin."
>
> "In fact," said Mustapha Mond, "you're claiming the right to be unhappy."
>
> "All right then," said the Savage defiantly, "I'm claiming the right to be unhappy."

Adrenaline-fueled fun is popularly attributed to stimulation derived from extreme sports such as BASE jumping (BASE is an acronym for the categories of objects jumped from: buildings, antennas, spans [such as bridges], and earth), free climbing on sheer cliffs without a rope, or extreme skiing on steep mountains prone to avalanches. Red Rock Canyon National Conservation Area in Nevada has some very challenging climbs; many climbers consider "Epinephrine" one of the best routes. Inexperienced climbers might have better luck playing roulette in nearby Las Vegas than gambling with "Epinephrine."

Dangerous occupations are also attractive to many people; for example, the headline of a CNN.com article on June 11, 2011, that described fighting a massive wildfire in Arizona that spread into New Mexico was "'Adrenaline Rush' Propels Fire Crews in Arizona." Working with fierce wild animals also qualifies. The late Steve Irwin, aka "The Crocodile Hunter," who participated in many documentaries, sometimes had a bold style with dangerous animals at close quarters. In response to a question about whether he might be perceived more as a thrill seeker than as a conservationist, Irwin responded: "I get called an adrenaline junkie every other minute. . . . You know what though, mate? I'm doing exactly what I've done from when I was a small boy. You can blame my dad for that, he started it. He created me. . . . Thrill seeker, adrenaline junkie? No problem at all. . . . I'm achieving, I am actually achieving conservation on a greater level than anyone thought possible, like Jacques Cousteau."[12]

Nevertheless, too much of any good thing can be harmful. Adrenaline Addicts Anonymous (AAA), a much younger organization than AA (alcoholics anonymous), suggests that "adrenaline is our drug of choice. We revel in coming as close to the edge as we can, without going off. We may even enjoy situations where we hazard everything. Yet we are rarely concerned about these risks. We feel that life without them would be too dull, too bland, too boring. Who could live like that? Not us!"[13]

Public Displays of Emotion

We are all familiar with emotionally induced visible color changes that can complicate some people's lives, such as blushing with intense embarrassment or blanching with anger. More subtle is the phenomenon of human body hairs standing on end as part of the fight-or-flight response. All these responses involve the autonomic nervous system. Much more dramatic are color changes in animals that may in part signal their inner life.

Aristotle noted the capacity of chameleons to change color. These changes occur within minutes in response to many environmental stimuli, including the presence of threatening predatory animals. The potential survival advantage of rapid color changes has been actively debated; color changes could play a role in sexual selection, control body temperature, conceal an animal in different environments, or serve as a display reflecting territoriality or danger. Chameleons are equipped with specialized cells called chromatophores that contain pigments that reflect colors. Clumped pigmented material in chromatophores has relatively little effect on the color of reflected light. However, dispersion of the pigment throughout these cells changes the color of reflected light.

In 1898, investigators demonstrated that adrenal medulla extracts modified the pigment in chromatophores in frogs. Subsequently, many scientists have demonstrated that adrenaline changes the dispersion of pigments in chromatophores in amphibians, fish, octopuses, and lizards.[14] Adrenaline also plays a role in bioluminescence. Some creatures, including fish as well as fireflies, can emit bursts of light as a result of chemical reactions in specialized organs. The light may have several purposes, including attracting potential mates. Midshipman fish of the Pacific coast of North America possess luminous organs that glow for hours after an injection of adrenaline. In 1932, the *Times* of London reported that Newton Harvey, a professor of biology at Princeton University and a leading authority on bioluminescence, had dredged up luminous fish from a depth of eight hundred fathoms near Bermuda and found that adrenaline injections lit up the luminous organs in this species. Injecting adrenaline into fireflies also turns on their glow. In 1926, the *New York Times* ran a story titled "Science Produces a Non-Stop Firefly," recounting research in which a firefly injected with adrenaline glowed continuously for twenty-four hours and then died—perhaps an early example of burnout. The light organ in fireflies is normally controlled by their nervous system using the neurotransmitter octopamine, a chemical structurally similar to adrenaline. Octopamine activates a specific receptor that signals the light to flash. This receptor has considerable homology with adrenergic receptors found in mammals and could be an evolutionary precursor of human adrenergic receptors.[15] Much more research is needed to more fully understand the significance of color changes in both land and sea creatures.

Does Adrenaline Contribute to Emotional Feelings?

A frequent physiological response to emotionally stimulating events includes secretion of adrenaline and activation of the sympathetic nervous system. Popular expressions such as "adrenaline rush" and "adrenaline-charged" suggest that elevated concentrations of adrenaline, or increased sympathetic nervous system activity, modify the emotional states that produced them. But can adrenaline's multiple physiological effects on the body really somehow change emotional responses in the brain?

Storytellers and scholars have written about human emotions for thousands of years. Ancient Greek physicians developed theories of human temperament based on the balance of the four primary humors (yellow bile, black bile, blood, and phlegm). However, scientific studies of emotions began in earnest only relatively recently. In 1872, Charles Darwin published a major book on emotions that provided an early framework based on observable phenomena.[16] In 1884, William James at Harvard University published a theory of emotions that proposed an essential connection between physiological changes in the body and emotions.[17] James claimed that the perceptions of insults or threats were cues that provoked responses in the body: when the brain received feedback about these changes in organ function, emotional states came to mind. In other words, he believed that feeling states in the brain did not initially trigger physiological changes in the body; rather, the feelings in the brain occurred because of changes in the body.[18] Carl Lange independently proposed a conception of emotions that overlapped with some of James's views.[19]

Their ideas have been lumped together as the James-Lange theory of emotions. The theory stimulated considerable criticism, as it did not have a solid experimental foundation. Moreover, the theory appeared counterintuitive, since the brain stimulated the organs of the body to respond in the first place.

Three decades later, Walter Cannon at Harvard Medical School forcefully challenged the James-Lange theory, even though earlier, as an undergraduate at Harvard, he had greatly enjoyed courses given by William James.[20] Cannon came to research on emotions through chance observations made while working on swallowing. As a medical student in 1896, Cannon worked with Henry Bowditch, who interested him in the use of X-rays to study previously invisible swallowing mechanisms.[21] Cannon noted that stomach and intestinal movement in experimental animals would suddenly stop from time to time. These interruptions, irritating at first, became interesting to Cannon when he noticed that the motility slowdowns occurred after the animals had been startled or became fearful or angry.

Years later, Cannon seriously pursued these observations in additional experiments that gave rise to his views on how the autonomic nervous system integrated organ function in response to urgent stimuli from the environment. He found that in threatening emergencies, animals augmented physiological responses essential for survival while temporarily shutting down nonessential functions such as churning food through the gastrointestinal tract. Cannon made major discoveries on the roles played by the autonomic nervous system and adrenaline in causing changes in the body that occur with intense emotions, drawing from data on the reactions of restrained cats placed close to barking dogs.

Cannon criticized James's theory on several grounds. Cannon believed that whether cats displayed fear or rage, the sympathetic nervous system was activated in exactly the same way. This result appeared in conflict with James's theory that different responses in the body would determine distinct emotional response in the brain. In addition, Cannon demonstrated that cutting the sympathetic nerves did not attenuate characteristic emotional patterns in cats, suggesting that the brain was not dependent on the sympathetic nervous system to either send or receive signals to express distinct emotions. In addition, Cannon argued that visceral responses occurred more slowly than the almost instantaneous outwardly manifested emotional responses; consequently, the brain could not be receiving information quickly enough from the body to explain rapid emotional responses. Furthermore, he found that artificially stimulating the sympathetic nervous system did not produce characteristic emotional responses in animals. Cannon interpreted this collection of experimental evidence as incompatible with the James-Lange theory of emotion.

However, Cannon's criticisms had important limitations. James addressed feelings in human beings, while Cannon focused on emotionally based behavioral responses in animals. Cannon's work in animals did not refute the possibility that the human brain could have changes in emotional feelings in response to the changes in the body. Using analytical techniques unavailable in Cannon's day, modern experiments have demonstrated that autonomic nervous system responses to emotional stimuli are not invariant, overturning one of Cannon's major objections to the James-Lange theory. In other words, stress responses are not as stereotyped as Cannon believed. In addition, we now know that these intense emotional responses stimulate the secretion of many chemicals, including the hormone cortisol from the adrenal cortex; these responses provide additional potential messengers for important feedback to the brain with signals unknown in Cannon's day. Consequently, these and other more recent findings diminish the force of his criticisms.[22]

The James-Lange theory has been tested by infusing adrenaline into volunteers to determine if this stimulates emotional reactions. In general, people infused with adrenaline are aware of changes in heart rate but do not assign emotional labels to their feelings. Some subjects experience watered-down emotions—for example, feeling "as if" they were afraid. Uncommonly, volunteers describe what seem like authentic emotional responses. In other words, the overall outcomes of these studies are ambiguous, neither strongly confirming nor refuting the James-Lange theory of emotions. Extrapolating from experiments conducted in sterile laboratory settings with relatively small doses of adrenaline cannot faithfully mimic the changes that occur with truly frightening or exciting events in the real world.

Experiments on the origin of emotions have been conducted in people with a severed spinal cord, since in these individuals sympathetic nervous system stimulation cannot reach the peripheral organs and most nerve signals from peripheral organs cannot reach the brain because of the block in the spinal cord. While there have been different outcomes in various laboratories, and experimental designs and methods have been criticized, taken together the results have not so far demonstrated compelling evidence of impaired emotional responses in these patients. Even though these results are compatible with the hypothesis that signaling the brain via autonomic or motor nerves is not important in determining emotions, these experiments are not definitive. People with a severed spinal cord retain normal nerve traffic through the cranial nerves, which do not traverse the spinal cord. In other words, these subjects' brains are not completely isolated from nerve feedback from the body. Indeed, limitations in investigating brains fully isolated from the body make it inherently difficult in principle to design an experiment that could refute the James-Lange theory.

We can be confident that fear triggered by the sudden appearance of a mugger stimulates brain centers to fire off signals that activate the autonomic nervous system and increase adrenaline concentrations in the blood. While adrenaline itself does not pierce the blood-brain barrier (a shield that protects the brain from many substances circulating in the blood), signals about the stressed state of the body reach the brain via nerves that take information from organs back to the brain and by changes in other chemicals in the blood that do cross the blood-brain barrier. Considerable interest has focused on the vagus nerve (cranial nerve X) and cortisol as candidates for transmitting information back to the brain that could modulate emotional responses. Conclusively determining the extent to which potential feedback mechanisms involving the brain actually influence feeling states will require additional creative research. In the meantime, adrenaline junkies will not be discouraged from getting their adrenaline flowing merely because we do not definitively know whether this response heightens their pleasure during stimulating adventures.

[APPENDIX]

Adrenaline's Nobel Connections: An Extended Cast of Characters

I can forgive Alfred Nobel for having invented dynamite, but only a fiend in human form could have invented the Nobel Prize.

ATTRIBUTED TO GEORGE BERNARD SHAW

Oliver and Schäfer's experiments with adrenal extracts provided enormous stimulation for further research. Over the past century, scientists have published several hundred thousand articles on adrenaline, related drugs, and the sympathetic nervous system. Exceptional scientists have the sophisticated judgment and capacity to select and solve the most significant and fundamental questions. Research on key problems involving adrenaline has led to discoveries of broad biomedical importance.

Alfred Nobel's estate established the Nobel Prizes, the *ne plus ultra* of awards for scientists. Nobel, a nineteenth-century Swedish chemist, made his fortune largely from dynamite, a safer form of nitroglycerine. Scandinavian committees meet in secret to select prizewinners; each year's laureates receive intense adulation and scrutiny. Some awards provoke controversy, especially when potentially deserving people are left out. Given the enormous size of the scientific enterprise, many great scientists will never receive this Prize.

Most of the fundamental discoveries involving adrenaline and noradrenaline that are intimately linked to Nobel Prizes are covered in Chapters 6, 7, and 8. These awards, shown in the Table, speak to adrenaline's overarching stature as a prototypical hormone and drug.

Nobel Laureates and Their Discoveries

1947	Carl and Gerty Cori	Glycogen metabolism
1957	Daniel Bovet	Antagonist drugs
1970	Julius Axelrod and Ulf von Euler	Neurotransmission in SNS
1971	Earl Sutherland	Hormone action and cAMP
1988	James Black	Antagonist drugs
1992	Edwin Krebs and Edmund Fischer	Reversible protein phosphorylation
1994	Alfred G. Gilman and Martin Rodbell	G proteins in signal transduction
2012	Robert Lefkowitz and Brian Kobilka	Studies of G-protein–coupled receptors

A remarkable number of additional Nobel laureates have adrenaline connections. Like the title character Zelig, the "human chameleon," in the Woody Allen film, adrenaline slips into many Nobel Prize pictures, running a gamut from tight scientific bonds to tangential associations. Some of the stories associated with these laureates illustrate the value of persistence, chance, and mentorship in science. Others provide a wealth of important, enlightening, or amusing stories. Laureates connected in some fashion to adrenaline are presented chronologically, except for several special-interest groups described near the end.

Additional Nobel Laureates Connected to Adrenaline

In 1920, August Krogh received the Nobel Prize in Medicine for discoveries involving the smallest blood vessels. Krogh studied the effects of adrenaline on these vessels.[1]

In 1932, Edgar Adrian and Charles Sherrington shared the Nobel Prize in Medicine for discoveries involving the functions of neurons.

Adrian conducted considerable research on neurons in the sympathetic nervous system. He studied at Trinity College, Cambridge, living for five years in rooms that once belonged to Isaac Newton. Adrian became the ninth Trinity person going to Stockholm "on the same errand." Sherrington witnessed some of the earliest dog experiments conducted by Oliver and Schäfer on adrenal extracts. Howard Florey (see below) conducted research in Sherrington's laboratory on the effect of adrenaline on blood vessels in the brain.

In 1934, George Whipple shared the Nobel Prize in Medicine with George Minot and William Murphy for a new treatment of pernicious anemia—a disease identified by Thomas Addison. In 1915, Whipple compared the ratio of adrenaline content in adrenal glands to the weight of the glands in patients and stressed dogs; he called this the "adrenalin index," but it had little utility. Later he discovered that liver supplements helped anemic dogs rebuild hemoglobin, which ultimately led to a treatment for an otherwise fatal disease.[2]

In 1937, Albert Szent-Györgyi received the Prize in Medicine for work on biological oxidization and identification of vitamin C. Szent-Györgyi's adrenaline connections include an attempt to identify a novel form of adrenaline in the adrenal medulla, a claim that Ulf von Euler later refuted.[3]

In 1938, Corneille Heymans received the Nobel Prize in Medicine for his discoveries involving chemical sensors that detect oxygen and regulate breathing. He made major contributions to understanding the sensors that detect changes in blood pressure and regulate sympathetic nervous system activity. Heymans published many papers on the effects of adrenaline on these sensors and the effects of impaired function of the sympathetic nervous system.

In 1943, George de Hevesy received the Nobel Prize in Chemistry for his innovative work with radioactive tracers. Hevesy studied with Fritz Haber (Nobel Prize in Chemistry in 1918) and Ernest

Rutherford (Nobel Prize in Chemistry in 1908). His methods using radioactive atoms to trace substances in the body opened many new vistas. In 1954 he published a paper on the effects of adrenaline on the distribution of radioactive phosphate in the body. He published extensively with the Nobel laureate Hans von Euler-Chelpin, father of Ulf von Euler.[4]

In 1944, Herbert Gasser and Joseph Erlanger received the Nobel Prize in Medicine for their discoveries involving single nerve fibers. Gasser worked with the Nobel laureate A. V. Hill on muscle physiology and with Henry Dale on acetylcholine and adrenaline responses in muscles. During World War I, Gasser and Erlanger investigated shock caused by traumatic injuries; many of their joint papers involved studies of the effects of adrenaline.

In 1945, Howard Florey shared the Nobel Prize in Medicine with Ernest Chain and Alexander Fleming for work on penicillin. Florey conducted research in Charles Sherrington's laboratory investigating regulation of blood flow in the brain. Using techniques developed by August Krogh, Florey found that arteries on the surface of brains did not constrict in response to adrenaline. Also, stimulating the sympathetic nerves going to the head did not visibly constrict cerebral arteries. These findings revealed the intricate control of the circulatory system: if blood vessels in the brain tightly constricted in response to adrenaline or noradrenaline, blood flow to the brain might be adversely diminished during stress responses. On the other hand, Florey found that stress in dogs caused intense constriction of blood vessels in the bowels, a useful regulatory response that temporarily diverts blood away from an organ not especially important during fight-or-flight conditions.[5]

In 1947, Bernardo Houssay received half of the Nobel Prize in Medicine for discoveries involving a hormone in the pituitary gland that influenced glucose metabolism. Houssay did considerable research with adrenaline including investigating control of secretion of adrenaline from the adrenal medulla, and its effects on glucose metabolism and smooth muscle. Houssay constructively advanced the debate on the vital importance of the adrenals.

In 1950, Edward Kendall shared the Nobel Prize for the discovery of cortisone, a major hormone of the adrenal cortex. During the Depression, attracting research funding was difficult. Kendall approached Parke-Davis with a proposal: in return for free adrenal glands, he would send back purified adrenaline equal to the amount Parke-Davis would have extracted themselves. Kendall needed only the cortex for his own research. The arrangement proved especially generous; Kendall's laboratory improved the yield of purified adrenaline, routinely isolating more than required to satisfy the agreement with Parke-Davis. Kendall sold the surplus adrenaline, generating additional research funds. From 1934 to 1949, Kendall's lab processed 150 tons of adrenal glands and delivered millions of dollars' worth of adrenaline to Parke-Davis. Kendall even purified adrenaline from a pheochromocytoma removed from a patient at the Mayo Clinic. Kendall demonstrated that adrenaline lowered the concentration of potassium in the blood. Injections of adrenaline analogs are now sometimes used in the treatment for life-threatening high potassium concentrations.[6]

In 1952, Selman Waksman won the Nobel Prize in Medicine for the discovery of streptomycin, the first effective anti-tuberculosis drug. Waksman worked in Jokichi Takamine's laboratory in Clifton, New Jersey, from 1918 to 1920; some of his work featured Takamine's diastase.

In 1963, John Eccles received a share of the Nobel Prize in Medicine for his discoveries involving transmission of signals between nerve cells. Eccles published a large paper in 1937 on the effects of adrenaline on the electrical properties of smooth muscle.[7]

In 1966, F. Peyton Rous received half of the Nobel Prize in Medicine for a discovery made fifty-five years earlier: he demonstrated in 1911 that an infectious agent caused cancer in chickens. The scientific community either severely criticized or ignored this discovery for a very long time. His papers involving adrenaline are connected to blood loss and inadequate tissue blood flow.[8]

In 1967, George Wald shared the Nobel Prize in Medicine for discoveries involving visual processes in the eye. Wald had studied with Nobel Prize winners Otto Warburg and Otto Meyerhof. Wald made many important discoveries about the function of rhodopsin, the light receptor in the eye; both α and β adrenergic receptors have considerable structural similarities to this protein. In 1952, Wald refuted a claim about the amount of adrenaline in mealworms, larvae of the beetle *Tenebrio molitor*. Wald spoke out as an early opponent of the Vietnam War in the 1960s; his name reportedly appeared on President Richard Nixon's "enemies list."

In 1968, Marshall Nirenberg and Gobind Khorana shared the Nobel Prize, along with Robert Holley, for coding of messenger RNA into amino acid sequences in proteins. Nirenberg later devoted his research to neurosciences. He investigated the drop in cAMP concentrations induced by adrenaline stimulation of α_2 receptors. Nirenberg found that with time cells became tolerant to adrenaline; cAMP concentrations gradually return to normal. In these cells, sudden withdrawal of adrenaline causes a dramatic rise in cAMP to values high above normal. Morphine and other opiate narcotics have similar biochemical effects in these cells and in the brain. Nirenberg's discoveries demonstrated important examples of how drugs can induce profound biochemical adaptations in cells. Al Gilman was a postdoctoral fellow in Nirenberg's laboratory. Khorana developed a major research program focusing on the relationships between the structure of rhodopsin and molecular mechanisms of visual signal transduction.

In 1973, Gerald Edelman shared the Nobel Prize in Medicine for work on the structure of antibodies. Later in his career Edelman took a profound interest in brain function. In 2004, some of his research involved investigation of noradrenaline as a neurotransmitter in the brain.

In 1976, Baruch Blumberg received a share of the Nobel Prize in Medicine for his work on the hepatitis B virus. Blumberg published a paper on adrenaline's effects on white blood cells in patients with Down's syndrome.

In 1977, Rosalyn Yalow received half the Nobel Prize in Medicine; Roger Guillemin and Andrew Schally shared the other half. Yalow received the prize for developing a novel assay for peptide hormones.[9] Guillemin and Schally identified hormones released by the brain to control the pituitary gland. Yalow utilized adrenaline in research projects both before and after developing this assay. Guillemin and Schally initially in collaboration and then independently—often in intense competition—provided many insights into how the brain influences hormone secretion by the pituitary gland. Guillemin and Schally separately purified and determined the chemical structures of several brain peptide hormones that increased pituitary hormone secretion.[10] Through the course of this work, both Schally and Guillemin published papers on adrenaline's capacity to influence hormone secretion from the pituitary.[11]

In 1980, Baruj Benacerraf shared the Nobel Prize in Medicine in 1980 for fundamental discoveries involving genes that determine immune responses. Benacerraf did research early in his career on the effects of adrenaline in various models of shock, and on animals after removal of the adrenal glands.[12]

In 1981, Torsten Wiesel shared the Nobel Prize in Medicine with David Hubel for work on the processing of information in the visual system. In 1974, Wiesel reported that dopamine rather than noradrenaline was likely a neurotransmitter in the retinas of octopus and squid eyes.

In 1982, Sune Bergström, Bengt Samuelsson, and John Vane shared the Nobel Prize in Medicine for their discoveries involving prostaglandins. Bergström initially demonstrated that prostaglandins were fatty acids and later solved the detailed chemical structure of many prostaglandins. Samuelsson did some of this work as a graduate student with Bergström and later made major discoveries about the metabolism of prostaglandins. John Vane discovered the major prostaglandin in blood vessels and solved how aspirin inhibited prostaglandin synthesis. Von Euler attracted Bergström to investigate prostaglandins. Bergström also collaborated with von Euler in the isolation of noradrenaline from the adrenal medulla. Samuelsson collaborated with Elias Corey at Harvard University. Corey won the Nobel Prize in Chemistry in 1990 for his enormous contributions to the synthesis of organic molecules including prostaglandins. Vane did extensive research on adrenaline. He worked in both academia and the drug industry. Like Henry Dale before him, friends tried to discourage Vane's move into industry.

In 1985, the Nobel Peace Prize was awarded to the organization International Physicians for the Prevention of Nuclear War. Bernard Lown, one of its founders, is a cardiologist who extensively studied the effects of adrenaline and the sympathetic nervous system on the heart. Yevgeny Chazov, the other founder, gave the Nobel Lecture: "We do not fully know the material basis of brain function, we do not know whether adrenaline, acetylcholine or opioid peptides determine human senses and behavioral reactions. However, we do know (such is, unfortunately, the nature of human consciousness) that most people, absorbed by anxieties of everyday life and with solving their daily problems tend to forget the global problems of life on Earth which concern all of us."[13]

In 1986, Rita Levi-Montalcini and Stanley Cohen received the 1986 Nobel Prize in Medicine for their multiple discoveries involving growth factors. They demonstrated that development of the autonomic nervous system in early life depends upon a hormone called nerve growth factor. Levi-Montalcini's interests focused on developmental biology; she discovered an unknown factor in tissues that stimulated nerve cell growth.[14] On the other hand, Cohen was interested in biochemistry; as a graduate student he investigated nitrogen metabolism in earthworms, spending many nights collecting worms for his experiments. Following his biochemistry fellowship, Cohen moved to Washington University and started working with Levi-Montalcini. They formed a wonderful collaborative team, bringing different skills together to work on the challenging problem of identifying the unknown factor. Cohen set out to purify it from tissue extracts. During the course of a many-year effort, Cohen attended seminars and lunches led by Arthur Kornberg. Kornberg suggested that Cohen add snake venom to his partially purified extracts to destroy contaminating DNA. Serendipitously, Cohen found that the snake venom itself contained a factor that vigorously stimulated nerve cell growth. Reasoning that snake venoms originate in modified salivary glands, Cohen made extracts from mouse salivary glands and happily found that they even more powerfully stimulated nerve cell growth than his original extracts. With this new source of highly enriched material, Cohen succeeded in demonstrating that the nerve growth factor was a small peptide. Levi-Montalcini and Cohen found that injecting newborn animals with antibodies that inactivate nerve growth factor prevented the development of a normal sympathetic nervous system, illustrating a major physiological role of this peptide.

Cohen noticed that injections of some partially purified extracts caused the closed eyes of baby mice to open earlier than usual. He decided to purify the factor in the extracts that had this odd effect and successfully identified a novel peptide called epidermal growth factor. An unpredictable dividend from this apparently esoteric research is that epidermal growth factor and its receptor play important roles in the growth of tumors. Blocking the effects of epidermal growth factor is now important in chemotherapy of lung cancer. This work illustrates how fundamental research can have unexpected dividends for human health.

In 1987, Jean-Marie Lehn shared the Nobel Prize in Chemistry for discoveries involving molecules interacting with high selectivity. Some of his work examined the binding of adrenaline and noradrenaline with novel synthetic macromolecules designed as chemical receptors. Lehn did postdoctoral research with Robert Woodward (Nobel Prize in Chemistry in 1965).

In 1997, Stanley Prusiner received the Nobel Prize in Medicine for discovering prions, infective agents responsible for bovine spongiform encephalopathy, aka mad cow disease. As a student, Prusiner investigated noradrenaline's effects on brown fat cells; this experience encouraged him to become a physician-scientist. His achievements support the notion that funding research positions for medical students is important in attracting physicians into research careers.

In 1998, Robert Furchgott, Louis Ignarro, and Ferid Murad shared the Nobel Prize in Medicine for discoveries about nitric oxide. Nitric oxide (distinct from nitrous oxide, aka laughing gas) increases the production of cGMP by the enzyme guanylyl cyclase. cGMP is a second messenger analogous to cAMP; for example, cGMP causes arterial dilation by activating a specific protein kinase. Sildenafil works in erectile dysfunction by inhibiting inactivation of cGMP. In 1978, Furchgott planned to reinvestigate β-receptor-mediated relaxation of blood vessels. Serendipitously, a technician's mistake led Furchgott to discover that acetylcholine dilates arteries indirectly by releasing a factor from the endothelial cells that line arteries. Ignarro provided evidence that the factor was nitric oxide. Murad had earlier demonstrated that the antiangina drug nitroglycerin worked by releasing nitric oxide. During World War II, Furchgott worked on severe shock, trying to isolate a factor that impaired adrenaline's capacity to raise blood pressure. Furchgott devoted much of his career to adrenergic receptor pharmacology. As a graduate student in pharmacology, Ignarro worked on noradrenaline in the sympathetic nervous system and later conducted research on many problems in adrenergic pharmacology including regulation of cAMP, and the actions of α and β receptors. Intent on going to medical school, Ferid Murad learned from Bill Sutherland, a fraternity brother, about his father's new MD-PhD program. Working in Earl Sutherland's laboratory, Murad made the discovery that adrenaline-stimulated cAMP accumulation occurred by the activation of β adrenergic receptors.

In 2000, Arvid Carlsson, Paul Greengard, and Eric Kandel shared the Nobel Prize in Medicine for discoveries in signal transduction in the nervous system: Carlsson for dopamine as neurotransmitter, Greengard for recognizing that many neurotransmitters activate protein phosphorylation cascades, and Kandel for mechanisms underlying learning and memory. Carlsson did research with Bernard Brodie at NIH on drug and neurotransmitter metabolism. Carlsson discovered that reserpine depleted noradrenaline from neurons.[15] He made many contributions to research on adrenaline and noradrenaline, including the development of important chemical methods for measuring catecholamines and their metabolites. In his role as a consultant to Hassel, Carlsson advised the company to seek β adrenergic receptor antagonists for the treatment of cardiac arrhythmias; he suggested DCI as the starting point for their work. The project led to the invention of metoprolol (see Chapter 9). Greengard did some research with H. Keffrer Hartline (Nobel Prize in Medicine in 1967) and postdoctoral work with Wilhelm Feldberg. He had extended visits in departments headed by the elder Alfred Gilman in New York and Earl Sutherland in Nashville. Greengard published many papers on the role of cAMP in protein phosphorylation, including how β receptors influence fundamental processes in cells. He studied noradrenaline release and signal transduction in adrenal chromaffin cells. His broader work on protein kinases has contributed to understanding signaling mediated via α_1 receptors. Kandel made a bold decision to adopt a simple model for his research on memory; he chose the giant marine snail (genus *Aplysia*) because its nervous system has relatively few neurons that are also very large. Choosing an appropriate experimental model often is a critical determinant of success. He focused on determining cellular and molecular events that contributed to learning and memory. He discovered that cAMP and protein kinase A in neurons play key roles in the snail's long-term memory.

In 2002, Robert Horvitz shared the Nobel Prize in Medicine for work on genes controlling cell death. He discovered that the programmed death of cells plays a key role in the development of organs. Horvitz did this research in a simple model, the worm *Caenorhabditis elegans*. He also studied the effects of octopamine, the worm equivalent of adrenaline. In his Nobel lecture Horvitz said, "I never expected to spend most of my life studying worms." He described the importance that childhood asthma had on his personal development, including a two-week hospital stay, "with an adrenaline inhalator next to my bed."

In 2003, Peter Agre and Roderick MacKinnon shared the Nobel Prize in Chemistry: Agre for work on channels that regulate the flow of water, and MacKinnon for discoveries on channels that regulate the flow of charged atoms (ions) in cells. Water channels in the kidney are involved in regulating urine volume. Agre discovered that phosphorylation of these channels by protein kinase A regulates their function. MacKinnon demonstrated that a specialized potassium channel is opened by cAMP, enhancing the flow of potassium out of cells. Most of the effects of cAMP are mediated by activation of protein kinase A; the direct effect of cAMP on this potassium channel represents a major exception.

In 2004, Richard Axel and Linda Buck received the Nobel Prize in Medicine for discoveries involving odorant receptors and the organization of the olfactory system. The molecular cloning of the odorant receptor genes owed much to the previously described DNA sequences of adrenergic receptor genes. Axel had earlier in his research career inserted the β_2 receptor gene into thyroid gland cells, giving the cells new capacity to respond to adrenaline analogs. As a high school basketball player, Axel had a formative experience: "I came onto the court as the starting center, and the center on the opposing team from Power Memorial High School lumbered out on the court, a lanky 7 foot 2 inch sixteen year old. When I was first passed the ball, he put his hands in front of my face, looked at me and asked, 'What are you going to do, Einstein?' I did rather little. He scored 54 points and I scored two. He was the young Lew Alcindor, later known as Karim Abdul Jabbar, who went on to be among the greatest basketball legends and I became a neurobiologist." After extensive experience in molecular biology, Axel devoted his efforts to brain research, quoting Woody Allen: "The brain is my second favorite organ." Buck has not published research directly involving adrenaline nor are her views on favorite organs readily available.

In 2008, Roger Tsien shared the Nobel Prize in Chemistry for work with green fluorescent protein. He shared the award with Osamu Shimomura, who had originally isolated green fluorescent protein from jellyfish, and with Martin Chalfie, who used the protein as a luminous genetic tag. Tsien engineered an assay that generates green light when cAMP concentrations change in living cells. Tsien also developed a dye to measure calcium concentrations in living cells, a method that has been used extensively to study the function of α_1 receptors.

In 2009, Elizabeth Blackburn shared the Nobel Prize in Medicine for discoveries involving telomeres and the enzyme telomerase. The ends of chromosomes are protected by special DNA sequences called telomeres that shorten with cell division; telomerase helps preserve telomere length. The shortening of telomeres has major implications for cellular aging and possibly in predicting mortality. Blackburn and collaborators have demonstrated that chronic psychological stress with increased concentrations of adrenaline is associated with decreased telomerase activity, shortened telomeres, and elevated risk of cardiovascular disease.

Surgeons and the Nobel Prize

Many of the earliest surgical laureates have adrenaline connections. In 1909, Emil Kocher received the Nobel Prize in Medicine for his work on the thyroid. In his Nobel lecture, Kocher described the interplay between adrenaline and thyroid function, including that adrenaline did not produce sugar in the urine of patients with hypothyroidism. We now know that β-receptor-mediated responses are diminished in hypothyroidism. Kocher was an early advocate for the use of adrenaline-containing adrenal extracts to treat cardiac arrest.

In 1912, Alexis Carrel received a Nobel Prize in Medicine for developing techniques to stitch blood vessels together and for work in organ transplantation. His prize was the second for a surgeon and the first for medical research done mainly in the United States. Carrel later developed techniques for keeping tissues and cells alive outside the body. He studied adrenaline obtained from Parke-Davis along with other ingredients in efforts to optimize preservation of organs.[16]

In 1923, Fredrick Banting became the third surgeon called to Stockholm; he shared with the physiologist John Macleod the Nobel Prize in Medicine for the discovery of insulin. Banting practiced orthopedic surgery and general medicine in a small Canadian city before starting his research on diabetes. Banting and Macleod used adrenaline to raise blood sugar in order to measure insulin's capacity to lower it. Banting recommended that if a large insulin dose caused hypoglycemia, adrenaline should be injected to ensure a prompt recovery. Macleod did considerable research on the metabolic effects of adrenaline prior to the discovery of insulin.[17]

In 1949, Walter Hess and Egas Moniz each received half the Nobel Prize in Medicine. Hess, originally an ophthalmologist, investigated how the brain controls the function of the autonomic nervous system. He discovered that specific parts of the brain regulated heart rate and blood pressure; activation of these centers mimicked responses to fear. His research is fundamental to the connection between emotional events and how the body reacts.[18] Moniz received his share of the prize for treating mental illness with prefrontal lobotomy, a surgical procedure that destroys part of the brain. The use of lobotomies illustrates a sadly recurring theme in medicine: new treatments can rapidly be applied to large numbers of patients in the absence of reliable knowledge about their benefits and adverse effects.

In 1956, Werner Forssmann (with nonsurgeons André Cournand and Dickinson Richards) shared the Nobel Prize in Medicine for work on heart catheterization. Forssmann did the first successful human cardiac catheterization by experimenting on himself, sliding a long catheter through an arm vein into his own heart. He walked to his hospital's radiology department, where an X-ray confirmed that the catheter had entered his heart. Forssmann received considerable criticism for his reckless and dangerous auto-experimentation. Nonetheless, Forssmann obtained authorization to catheterize a desperately ill patient to inject adrenaline and digitalis into her heart.[19] Cournand and Richards made major progress using cardiac catheterization in the 1940s. The technique has evolved into an essential diagnostic and therapeutic tool for cardiologists around the world. Richards did research with Henry Dale, studying the effects of adrenaline and the sympathetic nervous system on blood flow to the liver. Both Richards and Cournand were physicians who had broad interests in the regulation of blood flow by the autonomic nervous system.

In 1966, Charles Huggins received a share of the Nobel Prize in Medicine for the hormonal treatment of cancer. Huggins demonstrated that metastatic prostate cancer improved after patients' testicles were removed or after patients were treated with female sex hormones. He also developed hormonal treatment strategies for women with advanced breast cancer. In 1936, Huggins studied the effects of adrenaline on body temperature in humans and dogs. He also found that injecting adrenaline into the main leg artery effectively cut off blood flow; amputations performed after these injections hardly bled. He took note of Peyton Rous's much earlier research on adrenaline-induced arterial constriction. Huggins had a motto for his laboratory: "Discovery is our business."

The Cori Family Tree of Nobel Laureates

Carl and Gerty Cori had an extraordinary impact on many scientists; in addition to Earl Sutherland and Edwin Krebs, four subse-

quent Nobel laureates worked in their laboratories: Arthur Kornberg, Severo Ochoa, Luis Leloir, and Christian de Duve.[20]

In 1959, Kornberg and Severo Ochoa shared the Nobel Prize for
discoveries involving the biological synthesis of RNA and DNA.
Kornberg regarded the Coris as among his most important teachers;
moreover, their discovery of glycogen phosphorylase encouraged him
to seek enzymes responsible for DNA synthesis. Prior to joining the
Coris, Kornberg had worked in Ochoa's laboratory; Ochoa also contributed to Kornberg's passion for enzymes.[21] Ochoa worked with
the Coris on the metabolism of glucose-related molecules in the early
1940s. Ochoa had previously worked with two other Nobel laureate
scientists, Meyerhof in Germany and Dale in England.

In 1970, Luis Leloir won the Nobel Prize in Chemistry for work
on the biosynthesis of complex carbohydrates including glycogen. In
1944, Leloir left Argentina for a year to work with the Coris. Be-
forehand, Leloir had worked with Houssay on carbohydrate metabolism. Leloir collaborated on the discovery of peptides involved
in blood pressure regulation that contributed to major advances in
the treatment of hypertension. In accepting the prize, Leloir modestly paraphrased Winston Churchill's remark about British air-
men in the Battle of Britain: "Never have I received so much for so
little."

In 1974, Christian de Duve shared the Nobel Prize for discoveries
in cell structure. In his Nobel lecture, de Duve recognized the Coris
and Earl Sutherland as his early scientific mentors. De Duve had difficulty getting a position in the Cori laboratory; on his first application Carl Cori turned him down because "we do not see eye to eye
with respect to the action of insulin."[22] Cori later suggested that De
Duve work with Sutherland on the activation of glycogen phosphorylase. Afterward, De Duve worked independently on the problem;

however, Sutherland beat him to the punch with the discovery of cAMP.

How Sweet It Is: Adrenaline, Sugar, and the Nobel Prize

In the first fifty years of the twentieth century, a remarkable number of additional Nobel Prize winners worked on adrenaline and carbohydrate metabolism. Vincent du Vigneaud won the 1955 Nobel Prize for Chemistry, in part for synthesizing the peptide hormone oxytocin. He had earlier worked on the structure of insulin with John Jacob Abel. Du Vigneaud's work touched on adrenaline-induced hyperglycemia. In 1950, du Vigneaud demonstrated that the methyl group required to complete the synthesis of adrenaline from noradrenaline could come from the amino acid methionine. In 1953, Fritz Lipmann shared a Nobel Prize for discoveries involving the transfer of chemical energy during glucose metabolism. Lipmann published a paper using adrenaline in his study of glycogen synthesis. In 1915, Richard Willstätter received the Nobel Prize in Chemistry for research on plant pigments. He resigned his secure professorship in 1924 in Germany at the age of fifty-three on account of anti-Semitism involving a new faculty appointment. While he never set foot again in the University of Munich, students continued to work under his remote supervision. One of his final papers, published in 1936, investigated the action of adrenaline on glycogen breakdown. In 1938, Willstätter fled the Gestapo, reaching Switzerland, where he died during World War II. Albert Doisy shared the 1943 Nobel Prize for the discovery of vitamin K. He published on adrenaline and glucose metabolism. Otto Loewi, Fredrick Banting, and John Macleod did considerable research with adrenaline and glucose metabolism.

Works of Nobel Laureates in Literature that Mention Adrenaline

1925	George Bernard Shaw	*Saint Joan* ("diet of thyroid extract, adrenalin")
1928	Sigrid Undset	*Caterina av Siena*
1950	Bertrand Russell	*The Basic Writings of Bertrand Russell* ("both rage and fear, as we know from the work of Cannon, are due to secretion of adrenalin in the blood")
1962	John Steinbeck	*Sweet Thursday; In Touch; Once There Was a War* ("wolfed food distorts the metabolic pattern already distorted by the adrenalin and fatigue")
1976	Saul Bellow	*More Die of Heartbreak; Mr. Sammler's Planet* ("much adrenalin was passing with light, thin, frightening rapidity through his heart.")
1986	Wole Soyinka	*Seasons of Anomy* ("shot of adrenalin")
1991	Nadine Gordimer	*A World of Strangers; July's People* ("surge of adrenalin")
1992	Derek Walcott	*Conversations with Derek Walcott*
1993	Toni Morrison	*Paradise* ("feel the adrenaline")
2007	Doris Lessing	*The Sweetest Dream; The Fifth Child* ("surged with adrenalin")

Unrewarded Nobel Prize Nominees

Nobel Prize selection committees work in secret; however, the Nobel Foundation now provides access to names of nominees from more than fifty years ago. Walter Cannon received more than twenty-five nominations for the Nobel Prize, mainly for his work on the sympathetic nervous system and adrenaline. Some have suggested that Cannon might have joined Henry Dale and Otto Loewi in winning the Nobel Prize in Medicine in 1936 if not for his controversial theory involving multiple sympathins. Langley received more than ten nominations for his work on the autonomic nervous system. Schäfer received several nominations related to the discovery of adrenaline; his collaborator George Oliver received no nominations. John Jacob Abel received more than ten nominations for his work with hormones, particularly adrenaline; on the other hand, Takamine received no nominations for his contributions to the isolation of adrenaline.

Notes

I. THE GOLDILOCKS PRINCIPLE

1. A detailed account of the Gilbert case is given in William Phelps's *Perfect Poison*.

2. While Fränkel's paper is often considered the first clinical description of pheochromocytoma, a case report published about eighty-five years earlier in Ireland likely involved a fatal outcome from pheochromocytoma.

3. In 1900, Alfred Kohn found that chromium prominently stained the cells in the center of the adrenal gland where the adrenal medulla is located; he named them chromaffin cells. Pheochromocytomas are derived from chromaffin cells. While most chromaffin cells are found in the adrenal medulla, there is a relatively large collection of extra-adrenal chromaffin cells in the abdomen, near the division of the aorta into the two large arteries going to the legs. This group of chromaffin cells is called the organ of Zuckerkandl after their discoverer, Emil Zuckerkandl, who first identified these cells in human embryos, where they are relatively prominent compared to adults. About 10 percent of pheochromocytomas originate outside the adrenal medulla.

4. Unknown to Mayo, a few months earlier César Roux in Switzerland had successfully operated on a woman who had noticed a lump in her abdomen; the tumor was a pheochromocytoma in the right adrenal gland. Roux is famous for his innovative development of the Roux-en-Y anastomosis, used in major operative procedures on the gastrointestinal tract.

5. Myron Prinzmetal and colleagues conducted these experiments. As a student, Prinzmetal did early research with Gordon Alles on amphetamines (discussed in Chapter 9). Prinzmetal later demonstrated that some patients with angina developed chest pain due to spasms of the coronary arteries—called Prinzmetal angina—in the absence of atherosclerosis.

6. Serendipitous observations demonstrated that histamine injections provoked attacks of very high blood pressure in patients with pheochromocytomas by releasing adrenaline from the tumors. Other tests were developed that used injections of drugs that blocked some of the effects of adrenaline on arteries, lowering blood pressure in patients with pheochromocytoma. In both cases, the tests themselves could be risky, raising or lowering blood pressure to dangerous levels.

7. Genetic analysis in Roll's family members demonstrated mutations in the RET gene, known to cause pheochromocytoma along with other endocrine abnormalities, including thyroid gland tumors. Another inherited syndrome associated with pheochromocytomas is von Hippel–Lindau disease. In 2007, news reports suggested that von Hippel–Lindau disease occurs in members of the McCoy family, widely known for their deadly feuds with the Hatfield family beginning in the Civil War era. These stories speculated that the adrenaline from pheochromocytomas in some of the McCoys stimulated anger, exacerbating the feud.

8. The syndrome is named after Baron Münchhausen, an eighteenth-century German nobleman famous for telling untruthful tales.

9. Since the original description of this syndrome in 1991 in Japan, multiple reports have identified patients with similar findings around the world. In the medical literature, there are more than seventy different names for syndromes similar if not identical to takotsubo cardiomyopathy, including neurogenic stress syndrome and adrenergic cardiomyopathy.

10. Midodrine causes constriction of blood vessels by activating α adrenergic receptors, which normally are activated by the sympathetic nervous system. Midodrine has been used to treat postural hypotension since the early 1990s; it was approved by the FDA in an accelerated process aimed at expediting the availability of novel drugs that treat serious conditions that lack good therapeutic options. With this fast-track process, the usual stringent FDA standards are considerably relaxed, especially requirements for demonstrating major clinical benefits with adequate safety in large numbers of patients. In return, the FDA expects the company producing the drug to conduct further safety and efficacy studies after the drug is marketed. In August 2010, the FDA announced that it was initiating steps that could lead to removal of midodrine from the U.S. market because the company had not provided the FDA with compelling data demonstrating the drug's efficacy. This announcement stimulated an enormous outcry from both physicians and patients using the drug. An estimated 100,000 patients use midodrine in the United States; the press reported that many patients believed that the drug contributed to their living more normal lives. About a month later, the FDA indicated that it had no immediate plans to remove midodrine from the market; additional public meetings and reviews of data are planned.

11. Guttmann, a German Jew, left Germany in 1939 and moved to Oxford, England, where he went on to a distinguished career specializing in the care

of patients with spinal cord injury. He had a strong belief in the value of sports in the rehabilitation of his patients. In 1948, Guttmann organized the Stoke Mandeville Games, an athletic competition involving World War II veterans with spinal cord injuries. These games became increasingly popular under his leadership, with international competitions beginning in 1952. In 1960, the games were held in Rome following the Summer Olympics; these games are generally considered the first Paralympic Games.

2. RULED BY GLANDS

1. This anatomical region receives attention in Leviticus 3:3–4: "And he shall offer of the sacrifice of the peace offering an offering made by fire unto the Lord; . . . and the two kidneys, and the fat that is on them, which is by the flanks."

2. The history of the discovery of the adrenal glands and early ideas about their function are described in many sources, including Humphry Rolleston, *The Endocrine Organs in Health and Disease* (London: Oxford University Press, 1936) and Victor C. Medvei, *The History of Clinical Endocrinology* (Pearl River, NY: Parthenon, 1993).

3. Addison published in his book the full names of some of his patients. Increasingly stringent safeguards now mandate the protection of identifiable patient information in clinical research. Without these changes, the history of the United States in the second half of the twentieth century might have been very different. In 1955, a medical article describing the surgical management of patients with preexisting Addison's disease included the case of a thirty-seven-year-old man who had had a metal plate implanted in his back on October 21, 1954. The *New York Times* had reported on October 22, 1954, that Senator John F. Kennedy had had a spinal operation the day before, and on February 17, 1955, the *Times* reported that Senator Kennedy had had additional back surgery to remove a metal plate that had been inserted four months earlier. Putting these facts together would not constitute a two-pipe problem for Sherlock Holmes. Kennedy was frequently questioned about the health of his adrenal glands. Given his narrow victory over Nixon in the 1960 presidential campaign, knowledge that Kennedy had Addison's disease might have changed the outcome, even though he had been receiving completely effective therapy. Kennedy's autopsy confirmed the diagnosis of Addison's disease.

4. A famous early example of a finding at autopsy that explained the patient's symptoms is from 1723, in a case reported by Hermann Boerhaave. The grand admiral of the Dutch fleet engaged in three days of feasting followed

by an upset stomach that he tried to relieve by inducing vomiting. He then suddenly developed severe pain and survived only a short time. Boerhaave conducted an autopsy and found that the admiral's esophagus had ruptured—likely as a consequence of the vomiting. Rupture of the esophagus is now known as Boerhaave's syndrome.

5. Trousseau himself had several discoveries named after him, including Trousseau's syndrome, which describes the association between clotting in veins (especially in the arms) and cancer. Ironically, Trousseau developed a clot in a vein in his own arm and diagnosed himself as having cancer; he died from a stomach malignancy within months. Eponyms confer a form of medical immortality but are sometimes assigned to people who do not deserve recognition for a particular discovery. Indeed, Stephen Stigler proposed in 1980 a generalization he called Stigler's Law of Eponymy: *No scientific discovery is named after its original discoverer.* Interestingly, Stigler attributed the initial recognition of this law to Thomas Merton, so there is no logical conundrum arising from self-referral.

Many physicians now avoid using eponyms named after physicians complicit in Nazi medical atrocities. For example, arthritis and eye inflammation that follows from an infection elsewhere in the body was named Reiter's syndrome in honor of Hans Reiter, who reported a case during World War I. He joined the Nazi Party in 1931. During World II, he had involvement in medical experiments in Nazi concentration camps that led to the deaths of hundreds of prisoners. Based on an abysmal record that included complicity with involuntary sterilization, coupled with Reiter not being the first to describe this syndrome, there is an active, ongoing effort to replace "Reiter's syndrome" with the term "reactive arthritis."

Another controversy about eponyms involves whether to preserve the synthetic genitive form; namely, using the pattern of proper noun + apostrophe + *s* as in "Addison's disease." Dirckx has described a "war on the synthetic genitive" by zealous editorial reformers who prefer transforming proper names into adjectival forms, e.g., "Addisons disease."

6. J. M. D. Olmsted's *Charles-Edouard Brown-Séquard* (Baltimore: Johns Hopkins University Press, 1946) and Michael J. Aminoff's *Brown-Séquard* (New York: Oxford University Press, 2010) provide considerable information about Brown-Séquard's life and achievements. Originally known as Charles-Édouard Brown, he added his mother's name, emerging as Brown-Séquard in his earliest scientific publications, to "distinguish himself from all the other Browns."

7. Alfred Vulpian described a provocative Brown-Séquard experiment that involved cutting off a dog's head, waiting until all signs of movement disappeared, and then reconnecting the blood vessels to the brain. Brown-Séquard claimed that muscle movements returned to the dog's face. Vulpian suggested that Brown-Séquard should reconnect the vessels of the head of a guillotined prisoner, providing an opportunity to potentially lip-read thoughts after death.

   In 1865, Brown-Séquard became Harvard University's first Professor of the Physiology and Pathology of the Nervous System. He received no salary but could collect lecture fees from students and see private patients. The position was not a good fit, as he found the conditions for research unacceptable: "How can we learn to make scientific or practical investigations in physiology or medicine in a country like this, where the teaching of the sciences is yet only rudimentary? . . . [T]here is a great need here of an Institute, where the means of prosecuting scientific researches should be taught by competent men . . . to place this country on a level with Europe, for things about which the inferiority of America is notorious." C.-E. Brown-Séquard, "Medical Work and Medical Errors," *British Medical Journal* 1 (1867): 251–257.

8. In 1908, Edward Schäfer, co-discoverer of adrenaline, produced one of Harley's rats—now stuffed, and borrowed from a medical museum— during a lecture, to emphasize that Harley had been wrong.

9. More than two millennia earlier, Aristotle had familiarity with the effects of castrating animals. Castrating bulls to produce more docile oxen fostered agricultural development, and similar surgical procedures tamed aggressive stallions into calmer geldings or transformed the meat of a thin rooster into a plump capon.

10. Bernard's high standards and creativity are explored in Francisco Grande and Maurice B. Vissher, eds., *Claude Bernard and Experimental Medicine* (Cambridge, MA: Schenkman, 1967).

11. Incongruously, rumors pointed to Gull as the notorious serial killer Jack the Ripper, murderer of prostitutes in London in 1888. This bizarre, far-fetched story likely emerged from conspiracy theorists trying to connect the royal family with the murders.

12. Sophisticated programs are now in place to assess the quality of surgery, including risks of complications and death, for both hospitals as a whole and for individual surgeons. The number of patients who die within thirty days of surgery is a major outcome used to assess quality of care. Our

capacity to detect late-developing adverse effects of surgery and medical interventions including drug therapy is much more rudimentary, however. In the absence of effective monitoring systems, we remain dependent on astute clinicians and occasional prolonged clinical trials to identify unanticipated adverse outcomes. Epidemiologists also plumb the depths of huge databases containing clinical information on hundreds of thousands or millions of patients to identify unanticipated associations between therapeutic interventions and adverse outcomes.

13. Nearly a year before Murray's work was published in the *British Medical Journal*, Antonio-Maria Bettencourt-Rodrigues reported in Portugal that he had successfully treated a case of myxedema using injections of thyroid. His work was soon overshadowed by Murray's, however.

14. Rolleston's *The Endocrine Organs in Health and Disease* and Medvei's *History of Endocrinology* are excellent sources of information about thyroid disease in the nineteenth century.

3. A COUNTRY DOCTOR'S REMARKABLE DISCOVERY

1. Modern-day practicing clinicians, even in university hospitals, have great difficulty pursuing academic interests, let alone initiating bold research projects. Emphasis on continuity of care typically precludes prolonged absences from the clinic. However, Oliver did not have to contend with voluminous telephone calls, email messages, and insurance company billing forms.

2. Schäfer was English by birth, the son of a naturalized German merchant. After graduating in medicine from University College London, Schäfer's early research focused on microscopic examination of tissues (histology) and on the nervous system. After his collaboration with Oliver, Schäfer did considerable work in the emerging field of endocrinology. Schäfer rose through the ranks at University College to Jodrell Professor before moving in 1899 to become professor of physiology in Edinburgh. He developed a method of manual artificial respiration that was used for about forty years in cases of drowning. Both of Schäfer's sons were killed in action during World War I; his older son's name was John Sharpey Schäfer, in honor of his former teacher. After this son's death, to perpetuate Sharpey's name and in the context of anti-German sentiment during the Great War, Schäfer added Sharpey to his own name and dropped the umlaut from Schäfer, becoming Sharpey-Schafer. His grandson Edward Peter Sharpey-Schafer was later professor of medicine at St. Thomas's Hospital in

London; the younger Sharpey-Schafer did research in endocrinology and cardiovascular physiology that included experiments with adrenaline.

3. Oliver and Schäfer found that mixing an adrenal extract with stomach digestive juices did not diminish its capacity to raise blood pressure following subsequent injection into dogs. From this observation, many jumped to the erroneous conclusion that the substance would be well absorbed if given by mouth since it was not inactivated by stomach acids. Adrenaline is a good example of a drug that gets absorbed by the intestines but is inactivated in the liver before reaching the general circulation.

4. The Jagiellonian University, originally founded in 1364, educated Copernicus in the late fifteenth century and Pope John Paul II in the early twentieth century.

5. "Accident and Opportunism in Medical Research," *British Medical Journal* 2 (1948): 451–455.

6. "Present Condition of Our Knowledge regarding the Functions of the Suprarenal Capsules," *British Medical Journal* 1 (1908): 1277–1281.

7. H. Barcroft and J. F. Talbot ("Oliver and Schäfer's Discovery of the Cardiovascular Action of Suprarenal Extract," *Postgraduate Medical Journal* 44 (1968):6–8) explored the possibility that George Oliver used his son as an experimental subject. Oliver's son, Charles, became an engineer and inventor. Charles's two children did not recall their father ever indicating that he had been the subject of his father's experiments. Similarly, Edward Schäfer's daughter, a friend of Charles Oliver's sister, had not heard of George Oliver using his son in experiments.

8. *Endocrine* is derived from the Greek *endo*, "internal," and *krinein*, "to separate"; Edouard Laguesse coined the term in 1893. He studied clusters of cells in the pancreas that are distinct from cells that make digestive enzymes, naming them the islets of Langerhans in honor of the German scientist Paul Langerhans, who had discovered them about twenty-five years earlier while a medical student. Laguesse speculated that the islets secreted a substance that controlled blood sugar; this hormone was named *insulin* about twenty years later. Powerful modern techniques have demonstrated that the tiny islets of Langerhans have about five different cell types that produce several distinct hormones in addition to insulin.

Nicola Pende subsequently popularized the term *endocrine* in his writings. Pende, a professor of medicine in Italy, got caught up in repercussions from Benito Mussolini's manifesto that mandated segregation of Jews and other groups. To boost enthusiasm for the pseudoscientific basis of these policies, Mussolini demanded that prominent university faculty sign the

Manifesto. Apparently Pende objected before signing because he did not like the focus on superior Aryans at the expense of Italians.

9. "Sir Edward Sharpey-Schafer and His Contributions to Neurology," *Edinburgh Medical Journal* 42 (1935): 393–406.

10. Starling succeeded Schäfer as professor of physiology at University College London. Starling was an extraordinarily gifted and provocative scientist. He worked out the forces acting on blood in small blood vessels (capillaries) that helped explain the flow of fluids into and out of the blood. In addition, he discovered a key concept concerning the regulation of the pumping of the heart (Starling's law of the heart). He received nominations for the Nobel Prize in Medicine for his work on the hormone secretin. Bayliss was also a very distinguished physiologist. He married Starling's sister Gertrude. Later in his career, Bayliss was falsely accused of operating on a dog without anesthesia. He sued for libel in a legal action headlined in England as the "Brown Dog Case"; witnesses confirmed that the dog had been fully anesthetized, leading to a favorable verdict.

11. Pavlov confirmed these findings in his own laboratory, concluding: "Of course they are right. It is clear that we did not take out an exclusive patent for the discovery of the truth." Two years later, in 1904, Pavlov received the Nobel Prize in Medicine for his work on digestion. Pavlov later investigated conditioned reflexes, which included dogs salivating at the sound of a bell that they had previously learned to associate with the presentation of food.

12. The neologism *hormone* was likely suggested to Starling by the physiologist William Bate Hardy, who consulted the classical scholar W. T. Vesey.

4. FINDING A NEEDLE IN A HAYSTACK

1. Moore, an engineering graduate from Belfast, did research in Wilhelm Ostwald's chemistry laboratory in Leipzig before moving to Schäfer's laboratory in London to learn physiology. Ostwald was one of the great chemists of the nineteenth century, receiving the Nobel Prize in Chemistry in 1909. Ostwald's most famous students included Svante Arrhenius (Nobel Prize 1903), Jacobus van't Hoff (Nobel Prize 1901), and Walther Nernst (Nobel Prize 1920).

2. The songwriter/mathematician Tom Lehrer composed the song "Silent E," with the opening verse (used with permission from Mr. Lehrer):

Who can turn a can into a cane?
Who can turn a pan into a pane?
It's not too hard to see
It's silent *e*.

Changing *epinephrin* to *epinephrine* and, later, *adrenalin* to *adrenaline* are examples of an analogous process—adding a final *e*—but in these cases with no change in pronunciation.

3. The firm that later became Parke, Davis & Company originated in 1862 with Samuel Duffield, who sold compounded medicinal preparations to pharmacists. In 1866, Hervey Parke partnered with Duffield, forming Duffield, Parke & Company. A few years later A. F. Jennings bought out Duffield, leading to the firm Parke, Jennings & Company. When Jennings retired in 1871, Davis joined with Parke to form Parke, Davis & Company. With its research center in Detroit, the company operated manufacturing plants in the United States and internationally. Parke-Davis invested in scientific research in an effort to develop high-quality drugs and was once one of the world's largest pharmaceutical firms. Parke-Davis quickly recognized the importance of endocrine products, beginning to sell desiccated thyroid preparations as early as 1893 and adrenal gland preparations in 1895. (Parke-Davis also marketed a handy cocaine kit with syringe and needle before cocaine became illegal.) The company had several major vaccine programs and supplied large amounts of diphtheria antitoxin during epidemics in the early twentieth century. Parke-Davis remained independent until Warner-Lambert acquired the company in 1970; in turn, Pfizer acquired Warner-Lambert in 2000.

4. Before joining Parke-Davis in 1898, Aldrich had taught pharmacology with Abel at Johns Hopkins.

5. Stolz also synthesized noradrenaline at that time. Used in pharmacological experiments afterward, work went on for forty years before scientists realized that noradrenaline was a major neurotransmitter rather than merely a useful laboratory chemical.

6. Henry Dakin grew up in England and completed his undergraduate degree in 1901. In 1905 he accepted an unusual position in New York City, working with the physician-scientist Christian Herter in a biochemical laboratory Herter had installed in his mansion on Madison Avenue. He collaborated with Henry Dale on the immunological properties of albumin from both

chickens and ducks. From these sophisticated experiments Dakin liked to say that Dale and he had proved that a hen was not a duck. During World War I, Dakin worked with Alexis Carrel in a French field hospital and developed a very important antiseptic, Dakin's solution.

7. The story of the purification and synthesis of adrenaline is well told in the Davenport and Sneader articles found in the Further Reading section at the end of this volume.

8. Among others, Abel studied medicine with Adolph Kussmaul (Kussmaul's respiration in diabetic acidosis), Friedrich von Recklinghausen (von Recklinghausen's disease), Wilhelm Erb (Erb's palsy), and Max von Frey (von Frey hairs).

9. The dean of medicine at the University of Michigan asked Abel to recruit his own successor; Abel quickly identified twenty-seven-year-old Arthur Cushny, a Scottish pharmacologist, who agreed to assume the professorship. Years later, after returning to the United Kingdom, Cushny worked out the pharmacological activity of the isomers of adrenaline.

10. William de Bernier MacNider, "Biographical Memoir of John Jacob Abel (1857–1938)," *National Academy of Sciences Biographical Memoirs* 24 (1947): 231–257. Despite isolating an impure, chemically modified version of adrenaline, Abel's 1897 paper with Crawford is often viewed as the discovery moment for adrenaline.

11. Cited in K. K. Chen, ed., *The American Society for Pharmacology and Experimental Therapeutics, Incorporated: The First Sixty Years, 1908–1969* (Washington, DC: Judd & Detweiler, 1969).

12. ASPET annually recognizes outstanding research by young pharmacologists with the prestigious John Jacob Abel Award. Interestingly, from its inception ASPET barred from membership "persons in the permanent employ of a drug firm." Though this made many academics uncomfortable and led to some bitter protests, the restriction remained in place for thirty years.

13. The life and times of Abel are extensively discussed in John Parascandola, *The Development of American Pharmacology* (Baltimore: Johns Hopkins University Press, 1992). In 2008, one hundred years after its establishment, the American Society of Pharmacology and Experimental Therapeutics honored Abel as its founder. At that time, about five thousand scientists received an "Abel number," representing the closeness of their intellectual kinship with him. Someone who coauthored a paper with Abel was given an Abel number of 1. Publishing with an Abel coauthor merited an Abel number of 2, and so on. My own Abel number is 4.

14. The Tokyo Artificial Fertilizer Company passed through a series of mergers and takeovers and survives today as Nissan Chemical Industries.

15. *Koji* fermentation is also used to produce miso and soy sauce. Other species of *Aspergillus* have major industrial importance in synthesizing chemicals such as citric acid. Lovastatin, the first drug in its class approved to lower LDL cholesterol, is a metabolite found in another species of *Aspergillus*. On the other hand, *Aspergillus fumigatus* is a human pathogen that causes dangerous infections.

16. In 1833, Anselme Payen and Jean-François Persoz isolated a substance from a malt extract that converted starch into sugar; they named it *diastase* (from the Greek *diastasis*, meaning "making a breach"). This marked the initial discovery of an enzyme, though the term *enzyme* was not coined until years later.

17. Humans have exploited yeast, perhaps one of the first domesticated life forms, to make alcohol and bread for thousands of years. Eduard Buchner made the revolutionary discovery in 1897 that extracts of yeast could metabolize sugar into alcohol, demonstrating that this reaction could occur in the absence of living cells. For this work, he received the Nobel Prize in Chemistry in 1907. He had been a student of Adolf von Baeyer (Nobel Prize in Chemistry in 1905).

18. The *Chicago Daily Tribune* ran the headline "Whisky to Be Cheaper" over an article speculating that "there will be rejoicing in Kaintuck and Mizzoury when it is known that good whisky at 10 cents a drink is a probability of the near future." The Whiskey Trust was formed in 1887 and controlled a large number of distilleries, many of them in Peoria. This trust became one of the more notorious combines in US history, at one point dominating a large fraction of spirit production in the United States.

19. Manjiro Nakahama was one of the first Japanese visitors to the United States, arriving in 1843 in New Bedford, Massachusetts, after being rescued at sea by an American whaling boat. In 1850, Hikozo Hamada, also rescued at sea, later became the first Japanese naturalized American citizen. By the late 1860s, permanent Japanese immigrants and university-bound students began arriving in the United States. Many Japanese immigrated to Hawaii prior to its annexation as a territory by the United States in 1898. Thousands of Japanese took up residence on the West Coast of the United States, leading to strong anti-Japanese sentiments, especially in San Francisco. In the early 1900s, Japanese children endured segregation in the city's schools. In 1907, a "Gentleman's Agreement" between Japan

and the United States came into force. Japan agreed to limit immigration to the United States; in return, the U.S. government promised to improve conditions for Japanese already in the country.

20. One of the advantages of Taka-Diastase was its great potency; people had to swallow only small doses. Earlier malt extracts required bulky doses that people found objectionable.

21. Takamine and his son Jokichi Takamine Jr. applied for an early patent to use enzymes in cleaning clothes; the patent was awarded after the elder Takamine's death.

22. Nagai was sent from Japan to Germany as a member of a group of young graduates in the 1870s to learn about Western science and stayed in Germany for more than a decade doing research in organic chemistry. In Tokyo, Nagai purified ephedrine from ma huang, a medical herb used in China for more than five thousand years. He became a leading figure in chemical research on natural products and a noted leader in Japanese pharmacology.

23. Yoshizumi Tahara, a former student of Nagai's, worked in the National Hygienic Laboratory on the lethal chemical found in the liver and ovaries of fugu (puffer fish). Tahara used ammonia precipitation to partially purify the poison, now called tetrodotoxin. It is not known if this effective use of ammonia precipitation later gave Uenaka the idea to try it in the isolation of adrenaline.

24. Uenaka continued to work on products of interest to Takamine for several decades before returning to Japan, where he worked at the Sankyo Company.

25. Agnes de Mille, the famous dancer and choreographer, spent many summers as a child in Merriewold. De Mille is best remembered for her ballet *Rodeo* and for the choreography used in the Broadway musical *Oklahoma!*. Her uncle Cecil B. DeMille—he spelled the family name differently—was an Academy Award–winning Hollywood director and producer, responsible for the *Ten Commandments*. In her memoir about childhood, Agnes de Mille described extensive interactions with the Takamine family near their estate. Many people perceived the Takamines—who were wealthy and lived in an extraordinary Japanese-style house—as a royal family in residence. Diplomats and other distinguished guests visited on holidays. De Mille believed that people behaved respectfully in the Takamines' presence but that private conversations reflected concerns about the "Yellow Peril," along with concomitant worries about a Jewish invasion. She remembered Takamine speaking English with a Dutch accent and as a man

who enjoyed playing poker and fishing. De Mille also told an amusing story about her cousin, who got lost in a Kyoto department store while traveling with a group of students. She did not speak Japanese but uttered the two words she knew: "Jokichi Takamine." The manager of the store sent her back to her hotel in a taxi with a box of chocolates. This occurred nine years after Takamine's death; whether this gracious treatment reflects Takamine's fame or typical Japanese politeness remains uncertain. Several years after Takamine's death, his widow sold Sho-Foo-Den to John Moody, the founder of Moody's Investors Service. The house has subsequently changed hands several times.

26. Known as *Sakura* in Japan, these cherry trees have beautiful, short-lived blossoms that do not produce fruit. For many years, Eliza Scidmore envisioned the desirability of planting cherry trees in the Tidal Basin near the Potomac River in Washington; however, she made little progress moving the idea forward. Beginning as First Lady in 1909, Helen "Nellie" Taft took great interest in local civic improvement; she enthusiastically expedited a decision to plant cherry trees. Takamine learned of these plans but did not want to offend anyone by offering a private donation. Rather, Takamine made arrangements and paid, behind the scenes, for 2,000 trees to originate a gift offered by the Mayor of Tokyo. Unfortunately, this batch of trees was found to be infested with bugs and tree diseases on arrival in the United States. Takamine underwrote a second, even larger gift of trees, grown with intense oversight in Japan. These healthy trees were ultimately planted in Washington in 1912. Takamine also arranged for additional Japanese trees to be sent to New York City; many were planted at a location now known as Sakura Park. In Washington, some of the original trees came to a violent end after the bombing of Pearl Harbor in 1941. To diplomatically protect innocent trees, their names became "Oriental" or "Washington" cherry trees for the duration of the war. By some reports, the Japanese cherry trees at Parke-Davis, also a gift from Takamine in 1912, were cut down during World War II.

27. Two physicists with Riken connections later won Nobel Prizes: Hideki Yukawa in 1949 for his prediction of the existence of mesons, and Sin-Itiro Tomonaga in 1965 (shared with Julian Schwinger and Richard Feynman) for contributions to theoretical physics. However, Riken was not elitist and fostered industrial connections. For example, Ikeda Kikunae, a Riken scientist, identified and synthesized monosodium glutamate. Ernest Lawrence, mastermind of the giant cyclotron at the University of California,

Berkeley, provided advice as Riken began construction of its own cyclotron in 1938. Limited research on atomic weapons apparently took place at Riken during World War II. Riken houses the K Computer, designated in 2011 as the fastest supercomputer in the world then in operation.

28. For much more on Takamine's life and his scientific, diplomatic, and civic activities, see Mitsuo Ishida, *Jokichi Takamine* (Yokohama: Research Conference on Modern Creative Japanese Scientists, 2007); J. W. Bennett, "Adrenalin and Cherry Trees," *Modern Drug Discovery* 4 (2001): 47–51; and T. Yamashima, "Jokichi Takamine," *Journal of Medical Biography* 11 (2003): 95–102.

29. On the 150th anniversary of Takamine's birth in 2004, Japan issued a commemorative stamp showing his face with a superimposed representation of the adrenaline molecule.

30. Drugs often start out as a number, such as WB4101, assigned within a company or research institute to help keep track of substances as they are synthesized.

31. Takamine used both *adrenalin* and *adrenin* in patent applications. For several decades scientists used *adrenin* or *adrenine* interchangeably with *adrenaline*. Even Henry Dale sometimes used *adrenine* despite his argument with Henry Wellcome about *adrenalin*. Walter Cannon employed *adrenine* in multiple papers on the fight-or-flight response well into the 1930s.

32. Many drug companies offered their own versions of adrenaline under a host of trade names. Some examples include Adnephrin, Adrin, Atrabilin, Caprenalin, Chelafrinum, Epirenan, Haemostasin, Hemisine, Ischemin, Paraganglin, Paranephrin, Renoform, Supra-Capsulin, Supra-Nephran, Supra-Renaden, Supra-Renalin, Tonogen, and Vaso-Constrictin.

33. See E. M. Tansey, "What's in a Name?" *Medical History* 39 (1995): 459–476, and Jeffrey Aronson, "Where Name and Image Meet—The Argument for Adrenaline," *British Medical Journal* 320 (2000): 506–509.

34. Comprehending the expert testimony in the case is challenging without extensive knowledge of chemistry. Judge Hand's decision is both detailed and highly technical. He concluded his written decision with an expression of his displeasure with the way complex technical cases were decided in the United States:

> I cannot stop without calling attention to the extraordinary condition of the law which makes it possible for a man without any knowledge of even the rudiments of chemistry to pass upon such questions as these. The inordinate expense of time is the least of the resulting evils, for only a

trained chemist is really capable of passing upon such facts, e.g., in this case the chemical character of Von Fürth's so-called "zinc compound," or the presence of inactive organic substances. In Germany, where the national spirit eagerly seeks for all the assistance it can get from the whole range of human knowledge, they do quite differently. The court summons technical judges to whom technical questions are submitted and who can intelligently pass upon the issues without blindly groping among testimony upon matters wholly out of their ken. How long we shall continue to blunder along without the aid of unpartisan and authoritative scientific assistance in the administration of justice, no one knows; but all fair persons not conventionalized by provincial legal habits of mind ought, I should think, unite to effect some such advance.

Not much has happened to address Judge Hand's concerns, even though the science underlying many patent applications has only grown in complexity.

35. The patentability of natural substances remains a very active problem. For example, in 2012 a patent involving the human genes BRCA1 and BRCA2—involved in inherited breast and ovarian cancer—came before the U.S. Supreme Court. For a detailed discussion of the intricate legal aspects and implications of Parke-Davis's lawsuit against Mulford, see J. M. Harkness, "Dicta on Adrenalin(e)," *Journal of the Patent and Trademark Office Society* 93 (2011): 363–399, and C. Beauchamp, "The Pure Thoughts of Judge Hand," Brooklyn Law School, 2012, www.law.nyu. edu/ecm_dlv3/groups/public/@nyu_law_website__engelberg_center _on_innovation_law_and_policy/documents/documents/ecm_pro _071307.pdf, last accessed April 28, 2012.

5. ADRENALINE ZIPS FROM BENCH TO BEDSIDE

1. Hemophilia is an inherited bleeding disorder caused by deficiency in factor VIII, a key protein involved in blood clotting. In 1961, the British hematologist Ilsley Ingram demonstrated that adrenaline raised the concentration of factor VIII in the blood of normal subjects and in patients with mild or moderate hemophilia. This contributed to the discovery that a pituitary hormone called vasopressin increased factor VIII concentrations. Joseph Cort, working in Czechoslovakia, had synthesized analogs of vasopressin for entirely different reasons. One of these analogs, called DDAVP, later proved capable of substantially raising the concentration of factor

VIII. Clinical use of DDAVP enabled many hemophiliacs with mild to moderate bleeding problems to avoid use of factor VIII transfusions. Independence from these transfusions had an especially profound benefit in the period when the AIDS virus contaminated the blood supply.

2. Later in his career, Bates developed unique methods to improve vision. Believing that many visual complaints stemmed from eyestrain, he recommended a series of eye exercises for patients with various eye conditions. He also advocated gazing at the sun, a practice that can actually damage the retina. Bates's method is presented in his controversial book *Perfect Sight without Glasses*; his approach conflicted with scientifically based understanding of the visual system and was not generally accepted by other physicians. Nonetheless, Aldous Huxley believed that his long-standing severe visual dysfunction improved with the Bates methods and became a major advocate, especially in his book *The Art of Seeing.*

   In an odd footnote, in 1902 Bates disappeared from New York City and was found two months later working in London; he appeared half-starved but had not used any of his considerable savings. Several days later he disappeared again, only to be found by chance many years later practicing medicine in Grand Forks, North Dakota. He returned to New York City, where he again practiced as a physician. The explanation for these apparently profound memory lapses is unknown.

3. Parke-Davis sold adrenaline in many formulations: in solutions for injection under the skin, into muscles, or into veins; as an inhalant for administration via the lungs; as an ointment or cream for applying on the skin; as suppositories for administration via the rectum; or as tablets or lozenges for oral administration.

4. Anaphylaxis is a phenomenon known since ancient times. Around the turn of the twentieth century Charles Richet invented the word *anaphylaxis* to describe the sensitivity that could develop in some animals when reexposed to proteins or toxins. Richet received the Nobel Prize in Medicine in 1913 for his contributions to understanding these hypersensitivity reactions.

5. In 1910, a publication reported the possible benefit of adrenaline in patients with bubonic plague. Fifty consecutive patients received adrenaline treatment; the death rate in these patients was less than expected based on patients with the plague cared for earlier. Unfortunately, this is not a reliable method for concluding that a treatment is efficacious, especially since other aspects of patient care might have improved or the patients may not have been as sick.

6. Samuel Meltzer had a remarkable career as a physician and scientist. Born in Russia, he studied medicine at the University of Berlin, where as a student he did important research on swallowing, acting as one of his own experimental subjects. Soon after graduating in 1882, he moved to New York, where he practiced medicine while maintaining active research interests. He did research at night, both at home and in laboratories around the city before securing a full-time research position in 1904, in the Department of Experimental Physiology and Pharmacology at the Rockefeller Institute. He conducted considerable research on the actions of adrenaline, including early experiments on adrenaline and glucose metabolism.

   Meltzer also recognized the value of maintaining close intellectual contact between clinicians and laboratory-based investigators in selecting and solving key research problems. In 1907, at a meeting of the well-established Association of American Physicians, Meltzer recommended the formation of a new society to give younger physicians doing research the opportunity to present their work. These investigators typically did not have an opportunity to present their work at the Association of American Physicians (AAP) since its membership was limited to a small group of senior people. In 1908, the American Society of Clinical Investigation (ASCI) emerged. Successful election to membership required substantial research accomplishments. Members rallied to transform academic medicine in the United States into a discipline that had a strong research culture. At about the same time, Sultan Abdul Hamid II, the last Ottoman sultan with absolute power, was deposed in the aftermath of the Young Turk Revolution. The spirited members of the ASCI became known as "young Turks," while the more sedate, senior members of the AAP were viewed as "old farts." In 1940, the American Federation for Clinical Research was founded to give the most junior scientists opportunities to present their work; their members were soon known as "young squirts."

7. Trachoma is caused by the bacterium *Chlamydia trachomatis*. The disease has been essentially eliminated from the developed world but remains a major cause of blindness in the developing world.

8. Josef Mengele, the infamous World War II physician in the Nazi concentration camp Auschwitz, took a bizarre and futile interest in using adrenaline to change the color of the iris in the eyes of prisoners.

9. Crile made many ingenious contributions to surgery, including innovations in blood transfusion and in the design of surgical instruments. He later contributed to the founding of the Cleveland Clinic. In 1906, Crile performed

the first direct human-to-human blood transfusion. Despite its success, the practice of transfusion was not generally accepted until much later.

10. In his autobiography, Crile described how Arthur Cushny, the pharmacologist who replaced Abel in Michigan and would later identify the active isomer of adrenaline, gave him important encouragement to pursue his fundamental research at a time when surgeons were less enthusiastic. The eminent Norwegian Vilhelm Magnus wrote Crile, expressing great regard for his concepts involving surgical physiology and treatment of shock; Magnus lamented that most Norwegian surgeons only believed discoveries made in Germany. Crile ultimately obtained broad credit for his profound, life-long commitment to understanding the physiological changes that occurred during surgery.

11. In 1905, Crile collaborated with J. J. R. MacLeod (who in 1923 shared a Nobel Prize for his contributions to the discovery of insulin) on the limitations of adrenaline in the treatment of cardiac arrest caused by accidental electrocution. Nearly two decades later, Dr. Amos Squire, a physician at Sing Sing Prison in New York State, reported his belief that adrenaline would not revive prisoners who had been executed in the electric chair based on his studies of their isolated hearts removed at autopsy.

12. The treatment of animals with adrenaline was also newsworthy. In 1932, a gorilla at the National Zoological Park received adrenaline in an unsuccessful treatment for heart failure, and the *Washington Post* reported on April 6, 1940, that a team of veterinarians and physicians at Madison Square Garden in New York City had used adrenaline in treating Doushka, a Siberian leopard belonging to the Ringling Brothers–Barnum & Bailey Circus. The animal ultimately died from wounds suffered in a fight with his cage mate, an Indian leopard. The *Post* later reported the death of the gorilla N'Gi at the National Zoological Park. N'Gi received adrenaline for heart failure; his body was sent to Johns Hopkins University for further scientific study.

13. Solomon Solis-Cohen had a very distinguished career in academic medicine; he graduated from Jefferson Medical College and spent most of his career in Philadelphia. His family traced its ancestry back to the expulsion of Jews from Spain in 1492 during the Inquisition. He was a scholar who wrote many learned articles, translated poems from Hebrew, and published his own volume of poems.

14. Solis-Cohen suffered from hay fever and believed adrenal extracts were efficacious in relieving his own suffering. In light of the poor oral absorption of adrenaline, several writers have suggested that the favorable effects

Solis-Cohen reported in asthma were due to the as yet undiscovered adrenal cortical steroid hormone cortisol. We now know that high doses of steroids are effective in asthma; however, the extracts used by Solis-Cohen may have had too little cortisol to improve asthma.

15. Since adrenaline has a very short life span in the blood, patients required multiple daily injections. In the 1930s, scientists succeeded in slowing down the absorption rate of adrenaline into the blood after injections under the skin by combining adrenaline with peanut oil or gelatins. These ingenious maneuvers sustained the therapeutic benefit of adrenaline, permitting injections twelve hours apart. Other physicians tried using an electric current to increase the absorption of positively charged adrenaline through the skin. While this work had little clinical impact, the use of electricity to speed drug absorption—termed *iontophoresis*—has been reinvigorated by the modern challenge of administering novel peptide drugs that are not absorbed orally.

16. Interestingly, in 1942 the Council on Pharmacy and Chemistry of the American Medical Association noted its displeasure with the promotion and sale of adrenaline to members of the public. The modern blitz of promotion of potent drugs on television, in newspapers and magazines, and on the Internet by pharmaceutical firms represents a continuation of this practice.

17. Osler took a keen interest in Addison's disease early in his career. Reportedly, he had a patient with the disease at the Montreal General Hospital and anticipated examining the patient's adrenal glands at autopsy. However, after the patient's family declined permission for an autopsy, Osler slipped into the hospital's morgue the night before the funeral, thoroughly greased his arm, cut open the corpse's anal sphincter, pushed his arm through the bowel wall, and reached up to remove the adrenals without disturbing the outside of the body.

18. In the treatment of Addison's disease, many physicians in the early 1920s used a regimen that had been developed for a patient named Archibald Muirhead, a physician and professor of pharmacology who wrote an autobiographical note presenting the course of his own suffering with Addison's disease along with his response to treatment. The regimen involved as much adrenaline he could tolerate, administered either under the skin or rectally, along with an adrenal extract by mouth. Other patients received this regimen for at least another decade, although it likely had very limited effectiveness.

19. Despite modern emphasis on "evidence-based medicine," a substantial fraction of current diagnostic and therapeutic clinical practices lack a strong foundation.

20. Adrenaline emerged as the widely used initial treatment for anaphylaxis long before current standards of conducting randomized drug trials to establish therapeutic efficacy became commonplace. The uncertainty about the efficacy, dose, or most desirable route of administration of adrenaline in cases of anaphylaxis is highlighted by a 2009 Cochrane review of the efficacy of adrenaline in this condition. The Cochrane Collaboration is an independent group that systematically reviews the medical literature, digesting the best available information to inform medical decision making. This group could not identify even a single randomized trial that compared adrenaline to a placebo or to other sympathomimetic drugs. See A. Sheikh et al., "Adrenaline for the Treatment of Anaphylaxis: Cochrane Systematic Review," *Allergy* 64 (2009): 204–212.

21. The continued use of adrenaline in nasal surgery for more than a hundred years illustrates some of the challenges of determining a drug's benefits. Many surgeons inject adrenaline, cocaine, and a local anesthetic such as lidocaine into the septum in the nose before cutting, but few careful clinical trials have evaluated the benefits and risks of this triple combination. One 2007 carefully done study concluded that adding adrenaline to the two other drugs did not decrease bleeding or make operations technically easier. Since this study enrolled relatively few patients, a potential benefit of adrenaline might have been underestimated for statistical reasons, and it is therefore unlikely to transform surgical practice. Consequently, more than 100 years after adrenaline was first used in nasal surgery, we still do not definitely know if it is beneficial or potentially harmful.

22. J. Barr, "Paroxysmal Tachycardia," *British Medical Journal* ii (1904): 109–111.

23. In its report *To Err Is Human*, the Institute of Medicine estimates that as many as 100,000 hospital deaths per year are a result of preventable medical errors. Committee on Quality of Health Care in America, Institute of Medicine, *To Err Is Human: Building a Safer Health System* (Washington, D.C.: National Academies Press, 2000).

24. C. H. Burnett, "Results of a Mistake in Putting up a Prescription for Adrenalin Chloride to Be Used as a Nasal Spray," *International Clinics* 4 (1902): 25–26.

6. MIND THE GAP

1. Langley also recognized that the intestines had a separate nervous system—the so-called enteric nervous system—that controlled their function with considerable autonomy.

2. In 1905, Elliott published a comprehensive paper involving experiments in fifteen animal species, with measurements taken in at least nineteen different types of tissue. By contrast, in today's reductionist academic world, studying a couple of tissues from a mouse and perhaps a rabbit would represent very thorough animal experimentation, with many scientists preferring the simplicity of doing experiments only with purified molecules in test tubes.

3. In 1887, Emil du Bois-Reymond speculated that if neurotransmission was not electrical, then it was chemical. This broadly based conjecture came largely out of thin air, with nothing like Elliott's extensive empirical foundation.

4. Elliott was elected to fellowship in the Royal Society in 1913 at the age of thirty-five, based on the work he had done in Cambridge. After completing medical training, Elliott went on to a distinguished career as a professor of medicine in London; he had an enormous commitment to academic medicine with its missions in clinical care, teaching, and research.

5. Dixon, who taught pharmacology in Cambridge for many years, took a particular interest in psychoactive drugs, including cannabis, morphine, and mescaline, testing some of the substances on himself.

6. Sherlock Holmes first encountered Dr. Watson in a chemical laboratory at St. Bartholomew's Hospital, affectionately known as "Barts," as documented by Arthur Conan Doyle in *A Study in Scarlet*.

7. In 1925, Sinclair Lewis's novel *Arrowsmith* captured the sense that academic scientists had about a colleague who moved to a pharmaceutical firm: "Sorrowing men [scientists] wailed, 'How could old Max have gone over to that dammed pill-peddler? Why didn't he come to us? Oh, well, if he didn't want to—Voilà! He is dead.'"

8. The scientific name for the most prominent ergot is *Claviceps purpurea*. In the Middle Ages, ergot-contaminated rye caused gangrene due to severe arterial constriction in the limbs; the epidemics were sometimes called St. Anthony's Fire, after the early Christian ascetic St. Anthony, who reputedly helped several victims of this disease. Midwives used low doses of ergot to hasten childbirth, as ergot preparations can enhance contractions of

the uterus; this action is also useful after childbirth to combat postpartum hemorrhages.

9. Von Baeyer won the Nobel Prize in Chemistry in 1905 for work synthesizing organic chemical dyes.

10. Hunt received separate degrees simultaneously from two different institutions in Baltimore—a PhD from Johns Hopkins and an MD from the University of Maryland. He subsequently undertook several expeditions in Egypt, traveling far up the Nile to study the ancient *Polypterus*, a fish with lungs. Afterward, he worked on adrenal extracts and alerted American physicians to the dangers of methanol poisoning. His interests on structural analogs of choline likely stemmed from working with Paul Ehrlich in Germany from 1902 to 1904. He subsequently did research on acetylcholine at the Pharmacological Division of the Hygienic Laboratory, the precursor to NIH.

11. Dale discovered histamine in a sample of ergot, worked out its pharmacological actions, and elucidated the role of histamine in allergic responses, including life-threatening anaphylaxis. Dale found that extracts of the posterior pituitary gland caused vigorous contractions of the uterus, which led to the discovery of the hormone oxytocin. He played a major role in crafting international standards for the measuring the activity of drugs such as insulin.

    Dale later led the National Institute for Medical Research until his retirement in 1942. Among many other leadership positions, Dale was president of the Royal Society during World War II and served as chairman of the board of trustees of the foundation established by the estate of Henry Wellcome. He never held a primary university appointment. His daughter Alison married Alexander Todd, who won the Nobel Prize in Chemistry in 1957. Alison Dale published a research paper describing effects of adrenaline on blood flow in the lungs, citing a paper by her father.

12. O. Loewi, *The Workshop of Discoveries* (Lawrence: University Press of Kansas, 1953).

13. As a young scientist, Loewi visited England, where he met Henry Dale and Thomas Elliott in 1902. Some of Loewi's early experiments addressed adrenaline's effects on carbohydrate metabolism, an interest he sustained for more than forty years. In 1905, he published one of the very first papers demonstrating the pharmacological effects of chemically synthesized adrenaline and chemically related molecules prepared by Friedrich Stolz. When Hitler took over Austria in the Anschluss, Loewi was jailed for

several months; the Gestapo forced him to turn over his Nobel Prize money. He left Austria for England, initially staying with Henry Dale. He received support from the Society for the Protection of Science and Learning, which British academics helped establish in 1933 when the Nazis took power in Germany. Loewi obtained a temporary appointment in Belgium supported by the Francqui Foundation. Loewi happened to be in England on vacation when World War II broke out, precluding his return to Belgium. For a short time he worked in Oxford and then accepted a research professorship in pharmacology at New York University. He had some difficulties with his visa application to the United States. The U.S. consul in London asked Loewi to document his teaching experience; Loewi produced the *Who's Who* article that he had written about himself. Loewi spotted the physician's note on his medical certificate: "Senile, not able to earn his living." Fortunately, this was overlooked by the immigration official in New York. Loewi's wife, Guida, got clearance to leave Austria only in 1941 after transferring her family's property in Italy to the Nazis. She traveled in a "sealed train" from Berlin to Lisbon and then made her way to New York. The Loewis enjoyed academic life at New York University and at the Woods Hole Marine Biological Laboratory in Massachusetts, where they spent many summers. Guida Loewi died in 1958; Otto Loewi died at age eighty-nine and is buried in Woods Hole.

14. H. Dale, "II. The Beginnings and the Prospects of Neurohumoral Transmission," *Pharmacological Reviews* 6 (1954): 7–13. The phrase "egg of Columbus" alludes to criticism that arises when a clever discovery seems obvious after the fact. Christopher Columbus resented being told that any number of people could have discovered the Americas. He reportedly challenged some critics to stand an egg on end. When they all failed, he tapped one end of the egg, breaking the shell slightly, and stood the egg on that end.

15. Drugs that inhibit acetylcholinesterase, such as rivastigmine, are currently used in the treatment of Alzheimer's disease.

16. A year later, two German chemists found acetylcholine in the blood and other tissues of cattle. Dudley could not confirm their results, so he visited the German scientists in their laboratory, where he got the same results. However, upon returning to England he could no longer reproduce the experiments. Bruno Minz commented that the question of whether this discrepancy reflected differences between English and German scientists or cattle had not been solved.

17. Minz found temporary refuge in Paris and later escaped to Algeria. After the war he became a Professor in La Sorbonne in Paris.

18. In 1933, Krayer received an offer for the chair of pharmacology in Düsseldorf. Philipp Ellinger, the occupant of the chair, had been dismissed because he was Jewish. Krayer considered the offer unjust and declined. Consequently, the Nazi regime excluded him from holding a university chair in Germany. With the help of Henry Dale and A. V. Hill, and with support from the Rockefeller Foundation, Krayer got to England. He then spent three years on the faculty at the American University of Beirut in Lebanon. Afterward, he moved to the Department of Pharmacology at Harvard Medical School, later succeeding Ramsey Hunt as chair. Krayer made important contributions to the pharmacology of reserpine and its effects on the storage of noradrenaline.

19. Feldberg humorously described the importance of using the Hungarian leech, *Hirudo officinalis,* for the acetylcholine assay, claiming that the effects of physostigmine did not apply to "common English, German, or Israeli leeches."

20. Marthe Vogt's parents were leading neuroanatomists. She was a physician and had a PhD in chemistry from Berlin University. She obtained a major leadership position at a young age in Germany. While not Jewish, she decided to leave the country after Hitler gained power. She went to England in 1935 to work with Henry Dale. She was arrested as an "enemy alien" at the beginning of World War II. Dale and others helped expedite her release from custody. She continued to work on acetylcholine in the brain. In 1954, she described evidence suggesting that noradrenaline and adrenaline might be neurotransmitters in the brain.

21. Feldberg later demonstrated that acetylcholine stimulated the high voltage electric shock in torpedo rays. At the National Institute for Medical Research, he studied some of the effects of adrenaline on glucose concentrations in the blood.

22. Arturo Rosenblueth arrived in Boston from Mexico, supported by a Guggenheim fellowship. He was the eighth child of a Jewish Hungarian immigrant and a Mexican American mother. He played piano at a very high level and was deeply interested in mathematics and the philosophy of science. He received his medical degree from the National School of Medicine in Mexico City in 1927 after extensive study in Europe. The new Guggenheim fellowship program, aimed at fostering intellectual exchanges with Latin American countries, gave Rosenblueth the opportunity to

move to Harvard, where he encountered Cannon. Rosenblueth later collaborated with the MIT mathematician Norbert Weiner on research in cybernetics. Rosenblueth spent most of his later career doing physiological research in his native Mexico.

23. As it turns out, adrenaline is a major neurotransmitter in amphibians, including frogs; however, in the mammalian sympathetic nervous system noradrenaline is the primary neurotransmitter in adrenergic nerve endings. There are many species differences in autonomic nervous system function. Feldberg related how Dale encouraged him to do a physiological study on a lion's paw to check how closely it might resemble responses in domesticated cats; Feldberg, however, declined to enter its cage.

24. In 1937, investigators at McGill University suggested that noradrenaline had many of the properties of Cannon's sympathin-E. In 1938, scientists at Vanderbilt University reported pharmacological similarities between sympathin-E and noradrenaline. However, neither group offered evidence that noradrenaline existed in the body. In 1947, John Gaddum provided further support for noradrenaline being sympathin-E but did not use methods that allowed for chemical identification. Gaddum's sample of noradrenaline for comparison testing came from Blaschko, who had received an aliquot manufactured by IG Farbenindustrie in Germany just prior to World War II.

25. Peter Holtz graduated in medicine in Germany in 1928. He spent two years in England, first in Cambridge, and then with Henry Dale's group in London. Hermann Blaschko was also born in Germany. His father was a highly esteemed dermatologist noted for his public health efforts to prevent syphilis; his strong personal characteristics and humanitarianism influenced the quantum physicist Max Born. Decades later Blaschko supervised the research of Max Born's son Gustav in Cambridge. In 1924, Blaschko heard Otto Loewi's lecture on chemical transmitters in the heart, which stimulated his lifelong interest in this general area. The applied mathematician Richard Courant helped Blaschko obtain a position with Otto Meyerhof in 1925. Two years earlier Meyerhof had shared in the Nobel Prize with A. V. Hill. Blaschko later spent a year with A. V. Hill in London. Hans Krebs was Blaschko's physician during a hospitalization for tuberculosis in 1933; Krebs later received a Nobel Prize for work in biochemistry in 1953. Blaschko made many contributions to the study of adrenaline, including discoveries involving MAO and the synthesis of catecholamines.

26. Some postganglionic sympathetic neurons use dopamine as their principal neurotransmitter; dopamine primarily activates specific dopamine receptors analogous to α and β adrenergic receptors. Postganglionic neurons also release other substances as co-neurotransmitters, such as ATP and neuropeptide Y, but these have less importance than noradrenaline. Sympathetic neurons that activate some sweat glands use acetylcholine as their postganglionic neurotransmitter.

27. Ulf von Euler's father had worked with Walther Nernst (Nobel Prize in Chemistry in 1920 for his contributions to thermodynamics) and Svante Arrhenius (Nobel Prize in Chemistry in 1903 for his work on dissociation of electrolytes). Arrhenius predicted at the beginning of the twentieth century that accumulation of CO_2 in the atmosphere from human activities would lead to global warming. Arrhenius was Ulf von Euler's godfather. Von Euler's maternal grandfather was Per Teodor Cleve, a Swedish chemist and geologist who discovered several rare earth elements. At the age of seventeen, Ulf von Euler published his first scientific paper with his father. Leonard Euler, the superlative eighteenth-century mathematician, may have been a member of the same family. Among many breathtaking achievements in mathematics and physics, Leonard Euler derived an equation that Richard Feynman called the most remarkable formula in mathematics: $e^{i\pi} + 1 = 0$.

28. At Columbia University in New York, Raphael Kurzrok, an obstetrician and gynecologist, treated women with infertility by artificially implanting semen in the uterus. He noticed that the semen was frequently expelled from the uterus into the vagina. In collaboration with the pharmacologist Charles Lieb, he found that semen could variably cause contraction or relaxation of the uterus. Maurice Goldblatt, a lecturer at St. Thomas's Hospital Medical School in London, demonstrated that injections of extracts of human semen lowered blood pressure in animals.

29. After the war, von Euler urged the Swedish biochemist Sune Bergström to work out the chemical structure of prostaglandins. Bergström would share the 1982 Nobel Prize in Medicine with Bengt Samuelsson and John Vane for their discoveries involving prostaglandins.

30. Edith Bülbring was another German pharmacologist of Jewish ancestry who found work in England prior to World War II after intervention by Henry Dale. She initially assisted Harold Burn, chair of pharmacology at Oxford, and then evolved her own highly successful independent scientific career.

31. Erwin Chargaff was the senior author on a 1949 paper in *Science* that demonstrated the presence of noradrenaline in pharmaceutical-grade adrena-

line. He used paper chromatography to separate noradrenaline from adrenaline. Chargaff is best remembered for the "Chargaff Rules" about the chemical composition of DNA.

32. Patients with Parkinson's disease have too little dopamine in brain centers concerned with movement. These patients are frequently treated with large amounts of DOPA to increase the synthesis of dopamine in the brain.

33. The Catecholamine Club was founded in 1968 by prominent scientists broadly interested in adrenaline and related neurotransmitters. Julius Axelrod served as the club's first president. The official club song (sung to the tune of "Stout-Hearted Men") reflects those interests:

> Give me amines that are catecholamines that will bind to the membrane
> receptor.
> Start with tyrosine, dopa, dopamine and they'll soon give epi- and
> nor- for
> Synthesis and storage, release and reuptake take place as they 'complish
> their chore.
> When you need transmitters you can always count on them.
> When you walk with a shuffle and you can't move a muscle and your life
> has been
> filled with remorse,
> From the lab of Hornykiewicz it was clear that there was damage which
> shoulda been
> enough to kill a ho-orse.
> What you need is dopamine that's a catecholamine formed from levodopa,
> why of course.
> When you need transmitters you can always count on them.
> It can be inhibition that shatters your vision 'cause in war and in love we
> are blind.
> When your firing rates go wacko and your instincts run staccato
> excitations
> are the ones that come to mind.
> Plus and minus are both seen for each catecholamine in the cortex
> or in the brain stem.
> When you need transmitters you can always count on them.
> When you're fighting or shouting or downright knockouting put your
> faith
> in the adrenaline

Or in good vibrations. It might well be true sensations, you can trust what
receptors have seen.
Though you act like a teen due to catecholamine don't feel dizzy from collar to hem.
When you need transmitters you can always count on them!

*Oleh Hornykiewicz. In 1960 Hornykiewicz demonstrated that dopamine is depleted in the brains of people with Parkinson's disease and proposed an innovative treatment with the dopamine precursor DOPA, a drug that became a mainstay in the modern treatment of this disease. Several hundred neuroscientists published an open letter expressing disappointment that Hornykiewicz's contributions to the understanding and treatment of Parkinson's disease were not recognized in the Nobel Prize in Medicine in 2000, awarded, in part, for the discovery that Parkinson's disease is caused by a lack of dopamine in certain parts of the brain.

34. In the mid-1980s, investigators identified patients with an inherited deficiency of dopamine-β-hydroxylase. This rare disorder causes low blood pressure and other problems due to a lack of noradrenaline and adrenaline. A clever treatment is based on knowledge of the biosynthetic pathway of catecholamines. Dihydroxyphenylserine is metabolized by DOPA decarboxylase to noradrenaline; consequently, this drug can replace the missing noradrenaline in sympathetic neurons in patients with dopamine-β-hydroxylase deficiency. Blaschko had demonstrated the conversion of dihydroxyphenylserine to noradrenaline more than thirty years before the identification of this rare disorder. This novel therapeutic approach is a good example of fundamental research having unanticipated benefits for treatment of disease.

35. Henry Dale had discovered that injections of tyramine mimicked many of the effects of adrenaline. In 1928, Mary Hare (later Mary Hare-Bernheim) discovered in the liver a novel enzyme, tyramine oxidase, that inactivated tyramine. Several years later, Blaschko got a question from his boss, Joseph Barcroft: how does adrenaline get destroyed? Blaschko checked in the library and soon realized no one knew the answer; he decided to address Barcroft's inquiry in the laboratory. Blaschko and several collaborators identified an enzyme in liver that oxidized adrenaline. Their enzyme coincided with the enzyme isolated by Mary Hare years earlier. Almost simultaneously, Quastel identified the same enzyme while pursuing a different question in brain metabolism. Later named monoamine oxidase

(MAO), this enzyme proved important in the metabolism of adrenaline and noradrenaline.

Tyramine is contained in many foods; some cheeses and wines contain very large amounts. However, after absorption from the intestines, MAO in the liver immediately metabolizes the tyramine, preventing further penetration into the body. Without this protective mechanism, tyramine's sympathomimetic effects can cause life-threatening rises in blood pressure. Unfortunately, this problem came to light only after drugs that inhibit MAO activity became widely available to treat mental depression. Avoidance of tyramine-rich food became mandatory for these patients.

The discovery that MAO inhibitors are beneficial in depression came about through astute, unexpected observations made in clinical trials of a new drug for tuberculosis. Iproniazid produced undesirable psychic stimulation in many patients with tuberculosis. Could this adverse effect be transformed into a beneficial effect in patients with depression? In new clinical trials involving patients with depression rather than tuberculosis, iproniazid lifted symptoms of depression. Iproniazid also happened to be a MAO inhibitor in addition to having the capacity to kill tuberculosis germs. With this clue, pharmaceutical firms intentionally developed more powerful MAO inhibitors to treat depression. However, severe hypertensive crises occurred in some of these patients. These episodes appeared haphazardly, without apparent explanation. As it happened, in 1963, a pharmacist read a report by the British psychiatrist Barry Blackwell about symptomatic hypertension in patients taking MAO inhibitors. The pharmacist had noticed that his wife—prescribed an MAO inhibitor—developed similar symptoms after eating cheese. A letter from the pharmacist to Blackwell triggered recognition of the association between tyramine-rich foods and hypertensive responses in patients taking MAO drugs.

Opiate analgesics also have serious drug interactions with MAO inhibitors. Such an interaction may have occurred in the case of Libby Zion, an eighteen-year-old woman who died in a New York City hospital in 1985. She had been taking an MAO inhibitor and was treated with an interacting drug by a possibly sleepy resident physician late at night. Her death—perhaps from unrelated causes—precipitated a series of events that contributed to useful restrictions on the work hours of residents throughout the United States. Nonetheless, a more effective strategy for preventing this type of therapeutic problem is to widely implement electronic medical records with computer systems that instantaneously evaluate prescriptions for known drug interactions.

36. Dopamine in the brain is inactivated by COMT. Drugs that inhibit COMT are helpful in the treatment of Parkinson's disease by prolonging the survival of dopamine in the brain.

37. Axelrod's discovery of the reuptake mechanism occurred in the context of close interactions with clinical colleagues. Investigators at NIH began studies of catecholamine metabolism in patients with schizophrenia using radioactively labeled noradrenaline. Axelrod injected some precious tagged noradrenaline into animals; the noradrenaline disappeared quickly from the blood, taken up in their organs. With Georg Hertting, Axelrod cut the sympathetic nerves going to one eye in cats. After the postganglionic fibers had degenerated, these cats were injected with tagged noradrenaline; the noradrenaline appeared only in the eye with the intact nerves.

38. Bernard Brodie apparently got the nickname "Steve" after Steve Brodie, a New York bookie who may or may not have jumped off the Brooklyn Bridge in 1886, who became famous for surviving the fall. "Pulling a Brodie" meant doing something very bold. Bernard Brodie made extraordinary contributions to the understanding of drug metabolism; he won the Lasker Award in 1967 in recognition of those accomplishments. Brodie was born in England and grew up in Canada. He was an indifferent high school student, dropping out in 1926 and joining the Royal Canadian Signal Corps. After winning the Canadian Army boxing championship and with ample poker winnings to pay his way, he studied chemistry at McGill University. He later graduated with a PhD in organic chemistry from New York University. Brodie became interested in pharmacology and was influenced by Otto Loewi who had joined the faculty at New York University after escaping from Austria.

39. Axelrod wondered how amphetamine was metabolized in the body. He got general advice from Gordon Tomkins, who knew a lot about enzymes; Tomkins recommended doing experiments with isolated liver—the main site of metabolism—rather than in intact animals, which are too complicated. The tools needed for this type of research include an assay to measure the drug; an animal liver; and a razor blade to slice it. During the course of the work on amphetamine, Axelrod discovered an entirely new class of drug-metabolizing enzymes in the liver, now called the cytochrome P450 system. This discovery triggered a priority dispute with Brodie and stimulated Axelrod to become an independent investigator. To secure that type of position, Axelrod needed a doctorate. In 1955, at the age

of forty-two, he received a PhD from George Washington University. He discovered how melatonin, a chemical involved in daily sleep-wake cycles, is synthesized in the pineal gland in the brain. Axelrod's superb mentorship of young scientists produced many accomplished senior investigators. He is widely remembered for his special capacities to get to the heart of a scientific problem, for his creativity, and for his capacity to stimulate younger scientists.

7. HOW ADRENALINE STIMULATES CELLS

1. Ferdinand Blum remained a passionate investigator all his life. As a very senior scientist he explained his outlook: "Good Lord, what would I give to be 80 once more. What projects could I explore. But you know, at the age of 93 one must think about finishing one's life's work." He died a year later.

2. Carl Cori decided early on in favor of an academic career in medicine; his father had been a physician who did research in marine biology. At the height of the hippie movement in 1968 he reflected: "Suffice it to say that with a background of university professors on both sides of the family, it would have been unusual for me to go in a different direction. Rejection of the values of one's parents was not as prevalent then as it is today."

3. The Coris initially felt obligated to start investigation relevant to cancer. Several of their earliest publications in Buffalo addressed glucose metabolism in malignant tumors. Otto Warburg, who received a Nobel Prize in 1931, had demonstrated that cancers metabolized glucose much more rapidly than normal tissues. The Coris extended these results into intact animals. The typically high rate of glucose metabolism in many cancers is now exploited clinically with a radiolabeled glucose analog in positron emission tomography (PET) scans.

4. In 1922, as the Coris were getting started, the Nobel Prize in Medicine was divided between Otto Meyerhof and Archibald Vivian Hill. Meyerhof demonstrated that glycogen could be converted to lactic acid in muscle, and Hill was honored for his work on heat production in contracting muscle.

5. Carl Cori took an interest in the relevance of his animal research to human physiology. For example, he evaluated the effects of intravenous injections of adrenaline on blood sugar concentration in medical students.

6. Carl Cori had been warned that the chair of anatomy at Washington University was apparently not enthusiastic about his potential appointment to the faculty. While in the man's office for an interview, Cori casually mentioned that a bone on the desk evidently had come from the inner ear of a

whale. How he recognized the bone is unclear, but his comment ensured the chair's enthusiasm.

7. Gerty Cori was the third female Nobel laureate, preceded only by Marie Curie (who received a Nobel Prize in Physics in 1903, which she shared with her husband, and one in chemistry in 1911) and her daughter Irène Joliot-Curie (a Nobel Prize in Chemistry in 1935, shared with her husband). In 1979, twenty-two years after Gerty's death, the International Astronomical Union approved the naming of a crater on the moon in her honor—the Cori Crater, with coordinates 50.6S and 151.9W, and with a diameter of 65 kilometers.

8. In 1949, Linus Pauling discovered the first protein defect causing human disease: that the hemoglobin molecule is abnormal in sickle cell anemia. In 1952, the Coris presented the first evidence of a disease caused by an abnormal enzyme. About twenty years earlier, clinicians had described seriously ill children with abnormally low blood sugars that failed to rise in response to adrenaline; livers in these patients were enlarged and stuffed with glycogen. Gerty Cori focused on solving the molecular basis for a genetically recessive form of this disease. The Coris discovered that the livers lacked an enzyme needed to release glucose into the blood, causing glycogen to pile up. Stimulated by this seminal contribution, along with the added power of modern molecular genetics techniques, recent work has demonstrated many different types of glycogen storage diseases associated with a multiplicity of enzyme abnormalities.

   In 1951, Brian McArdle described another type of inherited glycogen storage disease characterized by exercise-induced muscle pain, stiffness, and weakness. McArdle's disease is caused by an enzyme defect in muscle that impairs glycogen breakdown. The patients' livers respond normally to adrenaline. The defect in McArdle's disease is due to inactivity of glycogen phosphorylase in muscle cells; however, the glycogen phosphorylase in liver, the product of a different gene, is entirely normal in these patients.

9. Sutherland recommended that stringent criteria must be satisfied before concluding that a hormone's effects were mediated by cAMP: the hormone must activate adenylyl cyclase, it must rapidly increase cAMP accumulation in cells, and drugs that inhibit the degradation of cAMP should enhance the hormone's effects. One final criterion—that cAMP must mimic the effects of the hormone—proved particularly challenging, since cAMP does not readily diffuse into cells. Theodore Posternak, a chemist visiting Sutherland's laboratory, synthesized cAMP analogs that got into intact cells relatively

quickly, and he demonstrated their capacity to mimic the actions of adrenaline on the liver.

10. The Carney complex—a rare inherited syndrome—is associated with abnormal PKA activity due to mutations in the gene for one of the regulatory subunits. This mutation may be responsible for the development of tumors in these patients.

11. In 2001, Leland Hartwell, Paul Nurse, and Timothy Hunt shared the Nobel Prize in Medicine for discoveries involving proteins that regulate the cell cycle, ensuring coordination between DNA synthesis and cell division. Many of these proteins are protein kinases.

12. Krebs recalled how he applied for one of his first grants: "Donald Hanahan . . . asked whether I had heard about National Institutes of Health research grants. He indicated that all one had to do to get one of them was to write a letter explaining what he or she proposed to work on. So I took an afternoon off, composed a letter, and in a few months I had my grant!" E. G. Krebs, "An Accidental Biochemist," *Annual Review of Biochemistry* 67 (1998): xiii–xxxii." At the present time, securing NIH grants involves a highly competitive process that requires detailed and laborious preparation of lengthy applications, with a success rate of 10 percent or less in many important research disciplines.

13. Arthur Kornberg published ten commandments in enzymology from his experience working on DNA replication. Commandment V stated: "Do not waste clean enzymes on dirty substrates." One of his students obtained a personalized California license plate that read PURIFY. Happily for Rodbell, later use of highly purified ATP confirmed the special importance of GTP.

14. Rodbell was influenced by discussions with Oscar Hechter about applications of cybernetic principles to endocrinology. In cybernetics, flow of information is fundamental, quite distinct from energy consumption. In his 1994 Nobel lecture, Rodbell describes how this framework came about:

> This subject was introduced to me by Oscar Hechter who had previously proposed several important theoretical considerations concerning hormone action. He was the first to question the proposition that hormones directly acted on the [adenylyl] cyclase enzyme. Through lengthy discussions at a downtown hotel bar in Washington, DC prior to a meeting that I had organized at NIH to honor Sutherland, we arrived at the concept of transduction as a means of coupling information between

signal-activated receptor and regulation of [adenylyl] cyclase. Given the paucity of knowledge at that time, the concept of informational processing was put in abstract cybernetic terms: discriminator for receptor, a transducer, and an amplifier representing [adenylyl] cyclase because of the large increase in cyclic AMP generated when converted to its activated state. The transducer is a coupling device designed to allow communication between discriminator and amplifier. (© The Nobel Foundation)

15. Gordon Tomkins was not only a brilliant scientist but also a person whose curiosity and imagination brought out the best in his colleagues. At a symposium in 1976 honoring Tomkins, Donald Fredrickson, director of NIH, fondly remembered Tomkins as a scientist and friend; see http://profiles. nlm.nih.gov/AA/Z/Z/Y/I/_/aazzyi.html (accessed March 22, 2010).

16. The latest edition of the book is edited by L. Brunton, B. Chabner, and B. Knollman: *Goodman and Gilman's The Pharmacological Basis of Therapeutics*, 12th ed. (New York: McGraw Hill Medical, 2011).The first edition of the book was an instant best seller. Years later, Al Gilman served as editor through several editions. The book's original purpose was aimed at educating medical students, emphasizing the integration of important information about drugs with medical sciences and therapeutics. Nonetheless, the book became the "bible" for professional pharmacologists. In the twenty-first century, the teaching of pharmacology to medical students has been abbreviated in many medical schools, with students relying on shorter books.

Goodman and Gilman made an important contribution to cancer therapeutics that is less well remembered. During World War II they worked on nitrogen mustard, which is similar to the mustard gases—sulfur replaced by nitrogen—that were used to devastating effect in World War I, for example at Ypres. The U.S. government funded classified research on the substance's pharmacology and toxicity with the goal of identifying novel antidotes. Early in 1942, Gilman, Goodman, and their colleagues demonstrated that nitrogen mustard killed many cells in the body; cells in lymph nodes were particularly susceptible. Extrapolating from this result, a mouse with cancer of lymphatic cells (lymphoma) was treated with nitrogen mustard; the tumor began regressing within days. This exciting result stimulated fundamental research as well as additional trials in animals. By the end of 1942, a human patient in the terminal stages of a lymphatic cancer received treatment with this dangerous chemical warfare substance. The clinical investigators had great courage in deciding on a dose that might kill cancer cells without killing the brave patient at the same time.

Remarkably, the patient had a very favorable initial response with considerable regression of the tumor. At that time, surgery and radiation were the mainstays of cancer therapy. The use of nitrogen mustard ushered in the era of medical oncology, the treatment of cancer with drugs.

17. M. E. Maguire, T. W. Sturgill, and A. G. Gilman, "Frustration and Adenylate Cyclase," *Metabolism* 24 (1975): 287–299.

18. Elliott Ross planned to do research on cyc⁻ cells with Gordon Tomkins in San Francisco after completing his PhD at Cornell University. However, Tomkins's death led Ross to Gilman's laboratory for his postdoctoral fellowship.

19. Gilman received an honorary doctoral degree from Yale University in 1997. The announcement of the honor quotes Gilman's response upon learning that he had won the Nobel Prize three years earlier: "First, I secreted all of my adrenaline; the adrenaline activated my receptors and they then stimulated my G proteins."

8. LOCK AND KEY

1. Ehrlich had many interests, ranging from microbiology and fundamental immunology to clinical medicine. He played a major role in discovering an early drug treatment for syphilis. Ehrlich shared a Nobel Prize in 1908 for work on immunity.

2. The Hill equation anticipated by a decade the derivation of a similar equation by Irving Langmuir to solve a different problem involving the adsorption of gases on a surface. Langmuir received the Nobel Prize in Chemistry in 1932 for his contributions to surface chemistry. Hill's work received little recognition at the time; he later studied the biophysics of hemoglobin, and muscle metabolism for which he shared a Nobel Prize.

3. Gaddum, a highly influential pharmacologist who had worked with Henry Dale, measured the capacity of adrenaline to contract rabbit uterus in the presence and absence of an antagonist ergot. Clark did analogous experiments using acetylcholine and the antagonist atropine. Clark had a long-standing interest in quantitative analysis; as a child prodigy, he did a calculation of the volume of Noah's ark that led him to doubt that even a pair of elephants could fit inside. In addition to his derivations of many important equations in receptor pharmacology, Gaddum participated in key experiments involving acetylcholine, the discovery that LSD antagonizes the effects of serotonin, and the discovery of substance P with U. S. von Euler.

4. Heinz Schild developed the Schild equation, which is useful in analyzing the effects of antagonist drugs. He worked in Henry Dale's laboratory in

England in 1932. Because he was a Jew, Schild decided to stay on in England after Hitler took power in Germany. As an Italian citizen, Schild was interned in 1940. Henry Dale, A. V. Hill, and others helped get him out and into war-related research.

5. Mark Nickerson worked out key properties of phenoxybenzamine, a drug that irreversibly blocks α adrenergic receptors. He won the Abel Award in pharmacology and wrote the chapters on adrenergic drugs in several editions of the Goodman and Gilman textbook. His academic career in the United States ended in 1954 during the McCarthy era. He was welcomed at the University of Manitoba in Winnipeg and later became chair of pharmacology at McGill University in Montreal.

6. Discovering that ergotoxine antagonized vasoconstriction mediated by adrenaline is a good example of serendipity in science, stemming from an error that Dale called a "shocking howler." At Burroughs-Wellcome, he was asked to test an adrenal extract for its adrenaline activity. He injected the extract several times into a cat; on each occasion the blood pressure fell rather than increased as expected. Dale reported that the extract did not contain adrenaline. He received a similar sample a week later and got the same results. Dale then realized that the cats in both experiments had received ergots during the course of an earlier experiment. He wondered if the preceding doses of ergot had somehow reversed the effects of adrenaline on blood pressure. Dale retested the suspect adrenal extract in a cat that had not received any other drugs; this cat's blood pressure rose dramatically. Dale concluded that the ergot preparation had transformed the effect of adrenaline on blood pressure. While Dale felt some embarrassment that his initial report had been wrong, he had made an important observation about ergotoxine that would only be fully understood years later. We now appreciate that the fall in blood pressure occurred because ergotoxine blocked α-receptor-mediated vasoconstriction and left untouched adrenaline's capacity to induce vasodilation mediated by β receptors.

7. Bovet initially did research on antibacterial sulfonamides. Gerhard Domagk at IG Farben in Germany demonstrated that the red dye sulfonamidochrysoidine (marketed as Prontosil Rubrum) was effective in treating bacterial infections. Domagk was awarded the Nobel Prize in Medicine in 1939 for this enormous discovery, but the Nazis prevented him from accepting the award at that time; he received it only after World War II. Bovet, working with Fourneau, discovered that sulfonamidochrysoidine's antibacterial activity involved metabolism in the body to the chemical sulfanilamide.

8. © The Nobel Foundation. Bovet invented the first antihistamine drug, now known as an H_1 histamine receptor antagonist. He went on to develop analogs of curare to relax skeletal muscle during surgery or paralyze patients requiring mechanical ventilation.

9. DCI has activity as a partial agonist: while blocking the actions of adrenaline, DCI partially activates β receptors on its own.

10. In 1772, William Heberden published his superb characterization of angina based on clinical observations. He described a strangling sensation in the chest that seized patients when they exerted themselves but which quickly disappeared with rest.

11. In retrospect, several curious precedents supported the potential efficacy of Black's general plan. Hypothyroidism decreases heart rate responses mediated by adrenaline. Deliberately causing hypothyroidism relieved severe, intractable angina. While causing one disease to treat another is not a desirable treatment, the benefits have heuristic value in support of Black's concept. Also, cutting the sympathetic nerves to the heart improved angina. However, many believed that the benefit from this treatment arose from cutting pain fibers leading from the heart back to the brain, making the patient dangerously unaware that his heart was deficient in oxygen. Wilhelm Raab made some remarkable observations supporting the idea that inhibiting the effects of adrenaline could improve angina. Raab irradiated the adrenal glands in a large number of patients with angina. He believed that the symptoms improved due to diminished secretion of adrenaline by the radiated adrenals.

12. ICI was taken over in 2008 by Akzo Nobel, an enormous Dutch multinational that itself had Nobel origins.

13. The Lilly program that synthesized DCI aimed at making long-acting β receptor agonists for the treatment of asthma; from that viewpoint, because DCI was a partial agonist it represented a backward step. Lilly apparently did not pursue development of β receptor antagonist drugs.

14. The German company Boehringer Ingelheim synthesized propranolol in a program developing drugs to suppress appetite. Recognizing the similarities between propranolol and DCI, their pharmacologists tested the capacity of propranolol to block the cardiac effects of an adrenaline analog. They obtained a positive result. However, not envisioning a clinical indication for propranolol, they did not pursue patent applications for this class of drugs until after ICI had established priority with its own applications.

15. Black's central idea for novel therapy for peptic ulcer disease focused on the fact that histamine stimulates acid production in the stomach. Antihistamine drugs used at that time for allergies did not attenuate acid production; these antiallergic drugs are now called H_1 histamine receptor antagonists.

    Over the past two decades, relentless biomedical advances have displaced H_2 receptor antagonists as mainstays for peptic ulcer disease. Novel proton pump inhibitors block the capacity of the stomach to make acid no matter the stimulus. Even more revolutionary has been the recognition that peptic ulcers are generally caused by the bacterium *Helicobacter pylori*, which can be eradicated with antibiotics.

16. Rhodopsin responds to photons by activating the G protein called transducin. Transducin stimulates biochemical changes that trigger the perception of light. In other words, the first steps in light perception are biochemically analogous to how cells respond to adrenaline.

17. A close reading of this list might raise a question about an apparently missing α_{1C} receptor gene. The internationally accepted terminology resolved some difficulties in the names given to α_1 receptor subtypes by different laboratories.

18. Specific parts of each adrenergic receptor determine how that receptor recognizes drugs and signals G proteins within cells. In Greek mythology, a Chimera is a fire-breathing monster composed of several animals—for example, a lion for the front, a goat in the middle, and a snake at the back for a tail. Engineering so-called chimeric receptors with parts from different adrenergic receptors provides a powerful technique for discerning the function of the various components of the native receptors and offering insights into human disease. In addition, Lefkowitz and collaborators demonstrated that changing a few amino acids in an α_1 receptor made it continuously active in the absence of adrenaline. As another example, a mutation in the gene for a G-protein-coupled receptor that stimulates testosterone production causes premature puberty in boys.

19. Lefkowitz was born in the Bronx, New York, and graduated from the medical school at Columbia University. While initially a clinical cardiologist by day and a scientist at night, he soon focused his energy on understanding the fundamental properties of the receptors for adrenaline and their implications for signal transduction mechanisms of G-protein-coupled receptors more broadly. As a postdoctoral fellow in his laboratory, I learned firsthand about his enormous enthusiasm, optimism, and passion for research, which happily infects young scientists who work with him.

Kobilka grew up in a small town in Minnesota and graduated from medical school at Yale University. As a postdoctoral fellow in Lefkowitz's laboratory, Kobilka made major progress in identifying genes for adrenergic receptors. After he moved to Stanford University, his fundamental research led to the first three-dimensional image of a β adrenergic receptor, a major step forward in understanding the interaction of G-protein-coupled receptors with drugs and hormones. During my time on the Stanford faculty, I recognized that Kobilka was not only an outstanding scientist but also a refreshingly modest person.

20. Thomas Sakmar at Rockefeller University has pointed out that the 7TM receptors are quite distinct from TM7, the nickname for John Sturges's film *The Magnificent Seven* (starring Steve McQueen, Yul Brynner, and other movie star subtypes).

21. The International Union of Basic and Clinical Pharmacology maintains a list of G-protein-coupled receptors, including detailed information about adrenaline receptors; see www.iuphar-db.org/index.jsp.

22. Robert Lefkowitz discovered that novel signaling molecules can attach to phosphorylated β receptors, leading to biological responses that do not depend on G proteins. The overall importance of these pathways remains a fertile field for further investigation and possible new drug development.

9. NEW DRUGS FROM OLD MOLECULES

1. High-throughput, automated assays now routinely screen thousands of chemicals for high-affinity interactions with cloned receptors; potentially interesting molecules can be identified in a matter of days. More intensive cellular and animal testing is still required to fully characterize activities of drug candidates, however. In addition, the likely metabolic fate of chemicals can be assessed with rapid methods using drug-metabolizing enzymes in test tubes.

2. More than a century ago, Paul Ehrlich emphasized the importance of the four Gs in drug discovery: *Geld* (money), *Geduld* (patience), *Geschick* (ability), and *Glück* (luck).

3. Several other discoveries contributed to the explanation of tyramine's actions. In 1931, J. Harold Burn demonstrated that destroying the sympathetic nerves going to the pupil obliterated tyramine's capacity to dilate the pupil; however, adrenaline continued to work under these conditions. Arvid Carlsson and colleagues discovered that tyramine did not raise blood pressure after treatment with reserpine. Carlsson had demonstrated in the 1950s that reserpine depleted noradrenaline from sympathetic nerve

endings. Several years later Burn and Michael Rand found that reserpine depleted noradrenaline from the walls of blood vessels; when they restored the noradrenaline, tyramine could again effectively constrict the arteries. These findings all suggested that tyramine worked indirectly, constricting blood vessels by releasing noradrenaline from sympathetic nerve endings in the arteries. Later experiments using radiolabeled noradrenaline directly demonstrated this.

In the twenty-first century, tyramine has revealed a new face to another generation of pharmacologists. Recent evidence suggests that small amounts of tyramine are synthesized by enzymes found in mammals, producing very low (trace) concentrations in the body. In 2001, scientists identified a novel family of human 7TM receptors, trace amine receptors. At least one of these trace amine receptors has very high affinity for tyramine. These receptors, found in the brain and other organs, are structurally related to adrenergic receptors. Amphetamine also interacts with these receptors, raising the possibility that amphetamine as well as endogenous substances such as tyramine may have novel, undiscovered effects mediated by these receptors.

4. The nasal mucosa has a complex structure, with venous tissues that can become quite swollen, much like erectile tissues in males.

5. While the agonist clonidine lowers blood pressure by stimulating α_2 adrenergic receptors in the brain, yohimbine (derived from the bark of *Pausinystalia yohimbe*) is an α_2 receptor antagonist that raises sympathetic nervous system activity and blood pressure by blocking these receptors. Purified yohimbine has been sold around the world as a reputed aphrodisiac and for erectile dysfunction, though there is only very weak evidence suggesting some efficacy in psychogenic impotence. On the other hand, yohimbine is FDA approved for reversing xylazine-induced sedation in free-ranging and confined members of the family *Cervidae* (deer and elk). Xylazine is an α_2 receptor-selective agonist that is frequently used to dart animals to induce temporary unconsciousness for management purposes.

6. Nagai was born in Japan and had an educational experience reminiscent of Takamine's formative expedition to the United Kingdom. After studying medicine, Nagai was sent in 1871 to Germany, where he did chemical research in Berlin for more than a decade. In 1883, Nagai returned to Japan as a professor of chemistry at the University of Tokyo, where he played a leading role in the development of the disciplines of chemistry and pharmacy in Japan.

7. In 1935, Bernard Read prepared a bibliography of ephedrine (cited in J. H. Gaddum, "The Alkaloid Ephedrine," *British Medical Journal* 1 (1938): 713–717).

Read's earliest citation for pharmacological experiments with ephedrine-containing medicines was in ancient China, followed by a gap of thousands of years. Between 1550 and 1880 there were merely several papers per century. The pace picked up to five to ten citations per year after Nagai's 1887 publication on the purification of ephedrine, only to explode into hundreds per year soon after Chen and his collaborator published their first paper in an English-language journal.

8. In 1925, Chen returned to the United States, where he continued research on ephedrine and subsequently received a medical degree. In later research, Chen purified chemicals from large quantities of toad venom; the venom contained adrenaline in addition to several novel chemicals related to digitalis. When Chen joined Eli Lilly in Indianapolis, he had to resign his membership in the American Society of Pharmacology and Experimental Therapeutics because industry scientists could not be members of the society at that time. Chen was later able to rejoin the society and became its president after attitudes towards industry scientists improved.

9. On account of its long safety record in pregnant women, ephedrine is a preferred drug to raise blood pressure if hypotension complicates spinal anesthesia during a Caesarian section. During spinal anesthesia, inhibition of the sympathetic nervous system can lead to marked falls in blood pressure, a worrisome complication potentially diminishing blood flow through the placenta to the fetus. Ephedrine has been used to treat men who fail to ejaculate sperm during orgasms. Normal ejaculation involves sympathetic nervous system-stimulated smooth muscle contractions in genital organs that propel seminal fluids containing sperm out of the penis. Ephedrine taken an hour before intercourse may enhance the movement of semen in the right direction during a male's climax.

10. While ephedrine is the major sympathomimetic drug in ma huang, the herb contains several other closely related chemicals. The two most prominent substances are pseudoephedrine and phenylpropanolamine (norephedrine). Pseudoephedrine is a sympathomimetic drug identified in the 1890s that remains available without prescription for nasal congestion. Phenylpropanolamine (aka PPA) was used for upper respiratory symptoms and obesity; it is no longer sold over the counter in the United States because of an association with increased risk of hemorrhagic stroke.

11. In 2003, evidence surfaced that American sprinter and broad jumper Carl Lewis had tested positive for ephedrine, pseudoephedrine, and phenylpropanolamine just before the 1988 U.S. Olympic trials. He had apparently

consumed a herbal supplement. While initially disqualified by the U.S. Olympic Committee, he received reinstatement because the use was considered inadvertent. Lewis finished second in the Olympic 100-meter race in Seoul; ironically, he was later awarded the gold medal when the race winner, Ben Johnson, was disqualified for testing positive for a banned anabolic steroid.

12. Clandestine chemists use ephedrine to simplify the synthesis of metamphetamine, a dangerously abused drug. Because their structures are so similar, only a few, albeit dangerous, chemical steps are needed to convert ephedrine to methamphetamine. Amateur chemists working in poorly ventilated home laboratories can cause fires, explosions, and produce lethal vapors.

13. Alles received a PhD from the California Institute of Technology in 1926. He interacted extensively with John J. Abel on the purification of insulin during Abel's extended visit to that institution. After graduation, Alles took a job in Los Angeles purifying pollens used in allergy testing. He interrupted this work for a postdoctoral year in Boston, where he did research with George Minot, who later won a share of the 1934 Nobel Prize for improving the therapy of pernicious anemia.

14. Lazar Edeleanu synthesized amphetamine in 1887, but its pharmacological properties were not understood until Alles conducted his experiments. Edeleanu later invented an important unrelated method used in petroleum distillation.

15. Toward the end of World War II, with penicillin in very short supply, Alles demonstrated that injecting penicillin intramuscularly with adrenaline prolonged penicillin's life in the blood. This modest effect likely involved delayed absorption of penicillin from muscles due to local vasoconstriction induced by adrenaline.

16. Narcolepsy is a disease characterized by chronic, excessive daytime sleepiness. Attention deficit/hyperactivity disorder is most often identified in children who have difficulty maintaining age-appropriate behaviors.

17. The distinction between β_1 and β_2 receptors contributed to the development of drugs for asthma with fewer adverse effects. On the other hand, the demarcation of β_3 receptors provided tantalizing ideas for novel drugs for obesity that have not so far come to fruition. This story illustrates the enormous hurdles—both biological and pharmaceutical—that must be overcome in making a successful new drug, particularly in extrapolating results in animal models to human disease. Considerable scientific talent sought novel sympathomimetic drugs for weight loss that specifically acti-

vate β_3 adrenergic receptors. The β_3 receptor subtype is highly expressed in fat. In obese rodents, β_3-receptor-selective agonists cause weight loss. Does the biology of activation of β_3 receptors in rodents provide a predictive model for treating human obesity? In rodents, β_3 receptor-selective agonists increase calorie consumption in brown fat by stimulating heat production. With very large ratios of body surface area to weight compared to humans, rodents rapidly lose heat and must burn many calories to maintain body temperature. Humans have comparatively small amounts of brown fat; consequently, if the weight loss in rodents depends on effects in brown fat, these drugs would likely have modest effects in people. Also, can the β_3 receptor selectivity of a candidate drug in rodents be extrapolated to humans? Rodents' β_3 receptors have a different protein structure compared to human receptors; consequently, a potent drug for the rodent β_3 receptor might not fit well into its human counterpart.

18. A sudden increase in asthma deaths was again noted in New Zealand in the late 1970s. Considerable evidence suggested that the introduction of the β_2-receptor-selective agonist fenoterol may have been responsible.

19. Recent evidence suggests that polymorphisms in the amino acid sequence of β_2 receptors influence the efficacy of these drugs in asthma. In particular, a single amino acid difference in the β_2 receptor protein (containing four hundred amino acids) can make a therapeutically important difference in drug responses. Further research in the relationship between β_2 receptor structure and response to drugs may help tailor therapy for individual asthmatics.

20. Ironically, prazosin emerged from a series of chemicals designed to slow the breakdown of cAMP in arteries. Despite this rational plan of drug development, prazosin actually lowers blood pressure by its unanticipated potency as an α_1 receptor antagonist.

21. A recurrent challenge in drug development involves the frequent reliance on surrogate end points in clinical trials. The goal of treating hypertension is to prevent complications such as heart failure. However, showing that an antihypertensive drug prevents heart failure or stroke requires vastly larger and more complicated studies than those needed to simply demonstrate that the drug lowers blood pressure—a convenient surrogate end point for therapeutic success. Achieving surrogate end points in hypertension (or for blood glucose in diabetes or for cholesterol in hyperlipidemia) may correlate well with meaningful clinical outcomes; however, for some drugs and diseases, surrogate end points may be misleading.

22. Prazosin and related α_1 receptor antagonists have an uncommon but problematic adverse effect: they can cause fainting by excessive lowering of blood pressure on standing—postural hypotension. Postural hypotension arises from the blockade of α_1 receptors in veins, preventing vasoconstriction. Normally, the sympathetic nervous system activates these receptors to prevent pooling of blood on standing. Many pharmaceutical companies sought novel α_1 receptor antagonists that potently blocked α_1 receptors in the prostate but had less capacity to block α_1 receptors in veins—a very demanding design for a new drug. Remarkably, the drug tamsulosin, discovered in Japan in the early 1980s, emerged as an α_1 receptor antagonist with these desirable characteristics. The explanation for tamsulosin's improved pharmacological effects rests on the biology of the α_1 receptor subtypes. Prazosin and related drugs have similar affinity for the α_{1A}, α_{1B}, and α_{1D} receptor subtypes. The pharmacology of the human body cooperated with the possibility of designing a better drug: α_{1A} receptors proved very important in mediating muscle contraction in the prostate, whereas α_{1B} receptors play a dominant role in human blood vessels including veins. Tamsulosin has modestly greater potency at α_{1A} receptors than at α_{1B} receptors; at an appropriate dose tamsulosin relieves symptoms of benign prostate hyperplasia without lowering blood pressure.

23. Metoprolol replaced practolol as a highly successful β_1-receptor-selective antagonist. Hässle—a subsidiary of the Astra drug company in Sweden—had synthesized metoprolol as part of its program to develop β receptor antagonists for heart rhythm disturbances worsened by adrenaline and sympathetic nervous system activity. Scientists at Hässle were unaware that ICI had already made enormous progress in its program aimed at inventing β receptor antagonists for angina. Metoprolol is currently one of the best-selling β receptor-antagonists in the world.

24. Brian Prichard conducted some of the earliest studies with propranolol in hypertension. His staunch advocacy for the benefits of propranolol in hypertension played a major role in persuading physicians that β adrenergic receptor antagonists are important drugs for this common medical problem. Prichard observed that overcoming resistance to this new therapy required not only excellent training in clinical research but also political campaigning experience, which he had received from his father, who was a politician in London. Prichard himself was elected the mayor of Wandsworth, England, in 2009, after forty years as a councilor.

25. In fiction, even lifesaving β receptor antagonists are not necessarily acclaimed miracle drugs. Henry Zuckerman, a thirty-nine-year-old man with coronary artery disease in Philip Roth's *The Counterlife*, stopped taking his β receptor antagonist because it interfered with his sex life and instead opted for coronary artery bypass surgery to treat his angina. The drug caused the known adverse effect of erectile dysfunction in this fictional character, making it impossible for him to service either his wife or Wendy, a very attractive dental assistant. Unfortunately, he did not survive the operation.

26. Propranolol was evaluated and approved by the FDA because of its efficacy and safety in treating angina. However, once a drug is marketed, physicians may prescribe medications for other clinical disorders. Off-label prescribing practices often represent first-rate medical care based on strong scientific evidence. While a pharmaceutical firm may subsequently seek FDA approval for additional indications for a drug, the work and costs involved may not make this worthwhile. However, without FDA approval for a specific indication, companies are not permitted to advertise this use of the drug.

27. *International Herald Tribune,*[o] October 21, 2004.

10. ADRENALINE JUNKIES

1. Nicklaus won the Masters by nine strokes that year; he held the record for lowest score and largest margin of victory until both were broken by Tiger Woods in 1997. A search in Google using the terms "Tiger Woods" and "adrenaline" generated about 1,990,000 hits in June 2012.

2. *World War Hulk Gamma Files*, Marvel Comics, 2007.

3. The effects of sympathomimetic drugs in muscle weakness have been exploited clinically due to a serendipitous discovery by Harriet Edgeworth, a patient with myasthenia gravis in whom ephedrine improved severe weakness. Myasthenia gravis causes profound muscle weakness; in adults this is generally due to antibodies directed against the patient's own acetylcholine receptors in skeletal muscle. In 1930, Edgeworth wrote in the *Journal of the American Medical Association* that her strength dramatically improved when her painful menstrual cramps were treated with ephedrine. She could rise from her wheelchair and even walk short distances. Physicians soon confirmed that ephedrine benefited other patients with myasthenia gravis. However, drugs that prolonged the effects of acetylcholine proved more

effective. Nonetheless, ephedrine remains useful in children with muscle weakness caused by a rare genetic mutation that impairs skeletal muscle contraction.

4. After more than two thousand miles of racing over three weeks, the very top finishers in the Tour de France are often tightly grouped—minutes or even seconds apart. A medication that slightly increases speed could change the outcome. Drug trials in patients typically aim to demonstrate clinically significant benefits, and often require hundreds or thousands of patients to demonstrate nontrivial changes in outcome. Setting aside ethical and recruitment challenges, a formal clinical trial to test whether or not clenbuterol could save mere minutes in the Tour would, in principle, represents an enormous challenge.

5. "Clinicopathologic Conference," *Johns Hopkins Medical Journal* 120 (1967): 186–199.

6. While some cases of being frightened to death are stranger than fiction, many stories and novels employ this concept. A good example is Sherlock Holmes's explanation for the death of Sir Charles Baskerville in the book *The Hound of the Baskervilles*, written by the physician-author Arthur Conan Doyle: "In that gloomy tunnel it must indeed have been a dreadful sight to see the huge black creature, with its flaming jaws and blazing eyes, bounding after its victim. He [Sir Charles] fell dead at the end of the alley from heart disease and terror."

7. The legal system appears to have accepted that emotional stress can be responsible for murder despite medical uncertainty. Multiple people have been convicted of homicide based on inducing emotional stress that was followed by the victim's death.

8. Some patients who have suffered a heart attack die from very slow heart rates due to excess activation of the parasympathetic nervous system. Curt Richter in the 1950s demonstrated that marked slowing of the heart was associated with death in some animal models of stress. "Storms" of either excessive sympathetic or parasympathetic activity can apparently be fatal.

9. Stimulation of sympathetic nervous system activity induced by vigorous exercise is also associated with the increased likelihood of rupturing susceptible plaques, leading to heart attacks and sudden death. A session of vigorous physical exertion is more likely to trigger a heart attack than spending the same time sitting in a chair, especially for people who do not exercise regularly. This may seem surprising because regular exercise decreases the overall risk of heart disease. The explanation for this apparently paradoxical

situation is that the likelihood of inducing a heart attack with any single episode of exercise is very small and is substantially outweighed by the beneficial effects of repeated exercise. Similarly, the risk of a heart attack during a sexual encounter is increased compared to simply going to sleep. However, the absolute risk of a heart attack with any single sexual event is very small; in other words, the vast majority of sexual experiences (or hostile emotional outbursts) are not followed by heart attacks.

10. Gustav Born, the son of the Nobel Prize–winning physicist Max Born, did important early work on the interactions between adrenaline and platelets.

11. A search in the Trademark Electronic Search System (TESS), the U.S. Patent and Trademark Office's online site, on June 7, 2011, yielded 351 records for *adrenaline* or *adrenalin*.

12. Sarah Simpson, "Interview with Crocodile Hunter Steve Irwin—Part 4: Adrenaline Junkie?," ScientificAmerican.com, March 25, 2001.

13. Quote from www.adrenalineaddicts.org.

14. Work with adrenaline stimulated considerable interest in the role of endocrine factors in skin pigmentation. Starting in the 1920s, Lancelot Hogben conducted experiments on the role of the pituitary gland in the change of skin color of frogs, leading to the discovery of a melanocyte-stimulating hormone that has a powerful capacity to darken skin.

15. Several Nobel laureates have had connections to octopamine. Paul Greengard played a major role in identifying octopamine receptors; Eric Kandel characterized an octopamine receptor in *Aplysia*, a large sea slug, his model organism for fundamental studies in memory. Julius Axelrod worked out sensitive methods for measuring octopamine in tissues. Robert Horvitz studied the effects of octopamine in his preferred model for developmental biology, the worm *Caenorhabditis elegans*.

16. While best known for his broad contributions to the theory of evolution, Darwin had a long-standing interest in emotions. He drew attention to the correspondence between emotional states in animals and humans, especially focusing on fundamental emotions such as fear and anger.

17. William James obtained a medical degree at Harvard Medical School in 1869; he later held senior faculty appointments in psychology and in philosophy at Harvard. James played a major role in defining psychology as a field distinct from both philosophy and physiology. He was the older brother of the novelist Henry James.

18. James described his view in the following way: "Our natural way of thinking about these standard emotions is that the mental perception of some

fact excites the mental affection called the emotion, and that this latter state of mind gives rise to the bodily expression. My thesis on the contrary is that the bodily changes follow directly the PERCEPTION of the exciting fact, and that our feeling of the same changes as they occur IS the emotion. Common sense says, we lose our fortune, are sorry and weep; we meet a bear, are frightened and run; we are insulted by a rival, are angry and strike. The hypothesis . . . says that this order of sequence is incorrect, that the one mental state is not immediately induced by the other, that the bodily manifestations must first be interposed between, and that the more rational statement is that we feel sorry because we cry, angry because we strike, afraid because we tremble, and not that we cry, strike, or tremble, because we are sorry, angry, or fearful, as the case may be. Without the bodily states following on the perception, the latter would be purely cognitive in form, pale, colourless, destitute of emotional warmth. We might then see the bear, and judge it best to run, receive the insult and deem it right to strike, but we could not actually feel afraid or angry." William James, "What Is an Emotion?" *Mind* 9 (1884): 188–205.

19. Carl Lange, a Danish physician, made important contributions to understanding mental illness, drawing attention to patients who suffered recurrent depressions. His brother Frederik Lange used lithium to treat hospitalized patients with depression, including patients with symptoms of bipolar disorder (manic-depression); with his death in 1907, the use of lithium for mental illnesses faded away, rediscovered by John Cade in Australia more than forty years later.

20. Cannon's excitement with philosophy led him to consider graduate work in that discipline rather than attend medical school. However, William James advised him: "Don't do it, you will be filling your belly with the East Wind." Cannon interpreted the remark as reflecting James's assessment that he lacked fitness; in any case, Cannon decided to follow the advice and went on to graduate in medicine from Harvard in 1900.

21. Henry Bowditch was a physiologist who helped Walter Cannon start original research as a medical student at Harvard. Bowditch had research training in Europe, including working in the laboratory of Claude Bernard in Paris. Cannon joined the faculty in the Department of Physiology at Harvard Medical School immediately after graduating; six years later he succeeded Bowditch as chair.

22. The brain centers that are important for emotional responses involve complex neural circuitry that is beyond the scope of the discussion in this book.

Several references to this research are provided in the Further Reading section, including material that discusses the importance of noradrenaline as a neurotransmitter in the brain.

APPENDIX

1. Krogh made many important contributions to the biology of very small blood vessels. In his Nobel Prize lecture, he calculated that if the total length of the smallest vessels (capillaries) in a typical man formed a continuous tube, it could go around the world at least twice. In 1922, shortly after the discovery of insulin in Canada, and motivated by the illness of his diabetic wife, Krogh played a major role in forming the company Nordisk to market insulin in his native Denmark. Although Krogh soon returned to his academic research, his highly successful venture is an early example of an academic scientist stimulating technology transfer to industry.

2. Whipple graduated in medicine from Johns Hopkins University, having taught pharmacology as an assistant in Abel's department. Several years later Whipple described a new disease—now known as Whipple's disease—in a single patient who had diarrhea, weight loss, and arthritis. Eighty-five years after Whipple's case report, the bacterium causing the disease was identified and named *Tropheryma whipplei* in his honor. Whipple studied anemia in dogs by making them anemic through repeated bleeding; he investigated how nutrition influenced rebuilding of new red blood cells in these dogs. George Minot in Boston had a long-standing interest in pernicious anemia. Since liver treatment in Whipple's anemia experiments was good for dogs, he wondered if it would help his patients. Minot and William Murphy evolved a special diet that included up to half a pound of liver daily for patients with pernicious anemia; many of the patients responded very favorably. Vitamin B_{12} proved to be the factor in liver that had such a favorable effect in pernicious anemia. Dorothy Hodgkin received the Nobel Prize in Chemistry in 1964 for determining the vitamin's highly complex structure.

3. The Hungarian-born scientist Szent-Györgyi was a highly dynamic investigator through his long career. In 1926, Fredrik Gowland Hopkins (Nobel Prize in Medicine in 1929) recruited Szent-Györgyi to the University of Cambridge. Szent-Györgyi crystallized an antioxidant from the adrenal cortex; since Szent-Györgyi had too little material to determine its structure, he reportedly decided to call it *ignose* (*ignosco* meant "I do not know" and *-ose* indicated that he thought it might be a sugar). However, the journal

in which he sought to publish rejected this name. His next choice, *godnose*, was also rejected. The name appearing in print was *hexuronic acid*. Szent-Györgyi secured an invitation to visit Edward Kendall's lab in Minnesota to gain access to large quantities of adrenal glands in the United States to purify much more hexuronic acid. After a year's work, Szent-Györgyi took the accumulated twenty-five grams of purified hexuronic acid back to his native Hungary. He readily purified large quantities in Hungary after discovering that paprika contains much hexuronic acid. Hexuronic acid proved to be vitamin C, the magic ingredient in citrus fruits that cures scurvy.

4. After Germany occupied Denmark in World War II, Niels Bohr (1922 Nobel Prize in Physics) worried about the security of the Nobel Prize medals of German physicists Max Von Laue (1914) and James Franck (1925), which had been sent to him for safekeeping. Transporting gold outside of Germany constituted a major crime under the Nazis. To ensure that identifiable medals would not be found in the event of a search, George de Hevesy dissolved the gold. The gold was recovered after the war, and the Nobel Foundation prepared new medals from it.

5. Florey, after working with Sherrington, later studied nerve-staining techniques with the Nobel laureate Ramon y Cajal in Madrid. Florey did some research on contraception; he demonstrated that copulation did not directly propel fluid into the uterus. There were some technical limitations in the study, as the dog volunteer did not copulate reliably or on schedule. In 1935, Florey became a professor at Oxford and recruited Ernst Chain, a refugee from Hitler's Germany. They became interested in isolating penicillin as a scientific problem. The clinical significance of the project accelerated during World War II.

6. As a graduate student Kendall worked on pancreatic amylase, an enzyme related to Takamine's diastase. Kendall later worked at Parke-Davis, doing research on thyroid hormone. He had to punch a time clock and disliked the industrial mind-set. Kendall moved to a hospital in New York, where he continued to purify thyroid extracts. He found the hospital's intellectual environment insufficiently stimulating; he moved to the Mayo Clinic, where he succeeded in isolating thyroid hormone and later isolated hormones from the adrenal cortex. A hundred years later, distinctions between intellectual environments in universities and biotechnology companies have markedly narrowed.

7. Eccles learned from the philosopher Karl Popper about the view that science grows through bold hypotheses that can be refuted or confirmed by

rigorous experiments. Apparently this philosophical outlook enhanced Eccles's own conceptual power and made it emotionally easier for him to accept the refutation of his own advocacy for the primacy of electrical rather than chemical transmission from nerves to target tissues.

8. Rous graduated in medicine from Johns Hopkins University in the same year as George Whipple. In 1911, working in New York, Rous reported that he had transmitted cancer by injecting cell-free extracts of a chicken's tumor into other chickens. The possibility that a virus was responsible for transmitting cancer in these experiments was not taken seriously: criticisms of the work included that Rous was not dealing with a true cancer or that he had inadvertently transferred intact cancer cells between chickens. After failing to establish a similar cancer model in mammals, Rous moved on to other research problems. During World War I, Rous developed methods that extended the shelf life of red blood cells preserved for transfusions. In 1934, Rous reentered cancer research, working with peculiar warts in wild rabbits that became malignant; in 1940 he established a transplantable tumor called V2. On account of the Nazi V-2 rockets later launched at Britain, he changed the name to VX2. Rous made many important discoveries involving the virus that caused this cancer, and also some involving chemically induced malignancies. About forty years after his initial discovery, the importance of the Rous sarcoma virus achieved broad recognition and acceptance. His daughter Marion married Alan Hodgkin, a professor at Cambridge University who shared a Nobel Prize in Medicine in 1963 for his work on how electrical signals move along the axons of neurons (three years before his much older father-in-law).

9. Yalow graduated from Hunter College in 1941 with a formidable academic record. Nonetheless, she had difficulty gaining admission to a physics graduate program; she ultimately received a PhD in nuclear physics from the University of Illinois in 1945. A few years later, Yalow joined the Veterans Administration hospital in the Bronx, contributing to the initiation of a clinical radioisotope program. She began to collaborate in research with Solomon Berson, a brilliant young physician. They discovered that insulin-treated diabetics had antibodies directed against insulin. They ingeniously used these antibodies and radioactively labeled insulin to develop a novel assay (radioimmunoassay) capable of detecting the very low concentrations of insulin in blood. Developing a practical assay involved overcoming major technical challenges. Yalow described in her Nobel lecture (and elsewhere) that their original paper on insulin-antibody interactions was rejected by

the very prestigious journal *Science,* and initially also by the well-known *Journal of Clinical Investigation,* as reviewers did not believe that insulin treatment of diabetics led to antibody production. Nonetheless, the latter journal ultimately agreed to publish the paper so long as the word *antibody* did not appear in the title. Success loves company; radioimmunoassays have since been developed to measure minute concentrations of hundreds of substances that are important in the daily care of patients, and used in all types of research. Yalow was the second American woman to win a Nobel Prize. Five years earlier, Yalow's research partner Solomon Berson died suddenly from a heart attack. Nobel Prizes are not awarded posthumously; in his presentation speech, Rolf Luft graciously spoke about the "Yalow-Berson" method. Yalow's successes reflect determination to overcome resistance, both as a woman in science and as a nonphysician doing research in a clinical department.

10. The pituitary gland secretes hormones that in turn stimulate the release of hormones from other endocrine glands, especially the thyroid, adrenal cortex, and gonads. In the 1930s, Francis Marshall proposed that the brain influences the pituitary gland. He demonstrated that electrical stimulation of the heads of female rabbits induced ovulation, an unusual experiment but provocatively compatible with this hypothesis. In the 1950s, Geoffrey Harris produced suggestive evidence that the brain nearest the pituitary gland—the hypothalamus—secreted unknown substances called "releasing factors" that enhanced the release of some pituitary hormones. Guillemin and Schally each used millions of hypothalamic brain fragments from tons of animal brains for their work, which took almost twenty years. They identified very small peptide hormones that stimulate the release of the pituitary's own hormones. The formidable obstacles in this biochemical tour de force cannot be overestimated.

11. Guillemin completed medical studies in France and moved to the University of Montreal to work with Hans Selye, the noted stress investigator. In 1953, he joined the faculty of Baylor University and later moved to the Salk Institute. Schally, born in Poland, survived the Holocaust as a teenager in Romania. He received undergraduate and graduate degrees from McGill University in Montreal. He worked in Murray Saffran's laboratory; their research strengthened the case that hypothalamic factors stimulated the release of pituitary hormones. Schally later joined Guillemin's laboratory at Baylor to work on purification of releasing factors. Schally next

opened an independent laboratory at the Veterans Administration hospital in New Orleans.

12. Benacerraf was born in Venezuela in 1920. His family soon moved to France, then left in 1939. Benacerraf received a medical degree in the United States at the end of World War II; he later practiced general medicine in the U.S. Army in France for several years. Intellectual curiosity led him to pursue research training in immunology, initially in New York and then in Paris. He returned to the United States to obtain independence in his research. As a professor at Harvard Medical School, a student reputedly asked him how to pronounce his first name. He reportedly replied, "You don't."

13. © The Nobel Foundation.

14. Levi-Montalcini grew up in Italy and graduated in medicine in 1936 with plans for further clinical and research training. Her experience during World War II and her emigration to the United States to do scientific research are inspirational stories. Mussolini's manifesto and laws that barred non-Aryan Italians from academic or professional careers blocked Levi-Montalcini because of her Jewish origins. She faced the dilemma of emigrating or dropping out of professional life in Italy. Levi-Montalcini decided to remain with her family in Italy and built a small laboratory in her bedroom. Research by Viktor Hamburger on embryological development in chicks inspired her. In 1941, she moved her laboratory from Turin because of Allied bombing raids. German occupation of Italy in 1943 forced her into hiding. She worked as a physician for refugees in Florence as the Allies moved north, and returned to academic life in Turin at the end of the war. In 1946, Hamburger invited her to join his research group in the United States. She sailed from Genoa, traveling with her medical school classmate Renato Dulbecco (Nobel Prize in Medicine in 1975 for work on tumor viruses). Another medical school colleague, Salvador Luria, shared a Nobel Prize in Medicine in 1969 for discovering viruses that infect bacteria. Levi-Montalcini was on the faculty at Washington University for many years before returning to Rome.

15. In addition to Carlsson, several other Nobel laureates worked with reserpine. George Palade shared the Nobel Prize in Medicine in 1974 for his discoveries in cell structure. He discovered granules in the heart and found that reserpine made the granules smaller; he speculated that they contained adrenaline. Others later demonstrated that the granules contained a novel peptide hormone secreted by the heart to regulate fluid balance.

Robert Woodward received the Nobel Prize in Chemistry in 1965 for synthesizing complex organic molecules. The presentation address indicated: "The synthesis of the famous poison strychnine caused a great sensation some ten years ago. Still more remarkable is perhaps the synthesis of reserpine, an alkaloid of great medical importance." © The Nobel Foundation.

16. Carrel was born in France and attended medical school at the University of Lyon. The assassination of Sadi Carnot, the president of France, aroused Carrel's interests in repairing blood vessels. Carnot died in Lyon in 1894 from a knife wound that cut the major vein going to his liver at a time when effective techniques did not exist for stitching blood vessels together. In 1903, Carrel accompanied patients on a pilgrimage to Lourdes. He wrote about a deathly ill girl who apparently recovered with prayers and holy water. Catholic clergy attacked him for expressing initial skepticism about faith healing, while physicians criticized him for gullibility. In 1904, after completing clinical training, he moved to North America, where he perfected his vascular techniques. In addition, Carrel claimed that his laboratory kept chick heart cells dividing in culture for decades; his ideas about immortal cells proved erroneous.

In 1935, Carrel collaborated with Charles Lindbergh to design a perfusion pump for heart surgery. Lindbergh's sister-in-law Elisabeth Morrow died of severe heart disease that could not be cured surgically in the absence of an artificial pump to circulate the blood while her heart was stopped for repairs. Carrel and Lindbergh made some progress with an innovative pump; however, cardiac bypass pumps became a reliable clinical reality only decades later. Lindbergh is best remembered for his solo airplane flight across the Atlantic Ocean in 1927 and for the kidnap-murder of his infant son in 1932.

Both Lindbergh and Carrel had favorable views of prewar Nazi-Germany. In 1935, Carrel published a best-selling book, *Man, the Unknown*, advocating selective human breeding and involuntary euthanasia. He supported the Vichy government until his death. Subsequently, Carrel received great recognition as a scientist in France; the medical school in Lyon and streets throughout the country were named in his honor. However, Carrel's political views suddenly reemerged from obscurity in the early 1990s when the ultra-right-wing National Front in France cited him in promoting their anti-immigration policies. The resulting controversy diminished his overall reputation; some Carrel street signs came down, and the Lyon medical school was renamed to honor René Laënnec, the French physician who invented the stethoscope.

17. Bitter controversy arose immediately after the Nobel Prize award to Banting and Macleod. Banting maintained that Macleod had not contributed to the discovery of insulin, whereas Charles Best, a medical student, deserved inclusion. Macleod noted the critical role of James Collip, who was instrumental in purifying insulin sufficiently enough for human use. Banting shared his prize money with Best, and Macleod did the same with Collip. Banting was very fortunate with insulin given his minimal scientific experience. He became very frustrated by his limited success in subsequent research projects. Banting died in an airplane crash during World War II. Despite early fame, Best wisely did further research training; he worked with Henry Dale on insulin-stimulated storage of glycogen in the liver. Best returned to Canada and had a highly successful academic career. Collip made major contributions to the purification of other hormones and worked on the development of Premarin, used for hormone replacement in postmenopausal women. Premarin contains a mixture of sex hormones to this day purified from the urine of pregnant mares.

18. Hess practiced ophthalmology and maintained a small research laboratory. His increasingly successful practice intruded on research, so he boldly gave up clinical work to study physiology. Hess's work provided an important foundation for psychosomatic medicine, a discipline concerned with mental phenomena affecting physical well-being.

19. Forssmann trained in surgery in Germany. After his dramatic initial human experiments, he did some animal research but soon focused on completing clinical training in urology. Forssmann joined the National Socialist Party (Nazis) in 1932. In 1937, Forssmann met Karl Gebhardt, Heinrich Himmler's personal physician. According to an article written years later by Forssmann's daughter, Gebhardt offered to recruit human subjects for Forssmann's research—an offer Forssmann declined. Gebhardt was hanged for war crimes and crimes against humanity after the Doctors' Trial in Nuremberg. Forssmann had difficulty finding employment in postwar Germany but managed in 1950 to start practicing urology in a small town; he was thrust into the limelight six years later.

20. Carl Cori tried unsuccessfully to recruit Hans Krebs to his laboratory in 1929. Hans Krebs would later share a Nobel Prize in 1953 for his biochemical discoveries.

21. Arthur Kornberg taught Paul Berg, who later shared a Nobel Prize in Chemistry in 1980 for his fundamental biochemical studies of nucleic acids, including recombinant DNA. Arthur Kornberg's son Roger Kornberg

won a Nobel Prize in 2006 for work on the molecular basis of the transcription of RNA from DNA. Roger Kornberg had research experience as a high school student in the laboratory of Paul Berg. His first scientific paper included as coauthors the Nobel laureates Paul Berg and Arthur Kornberg, as well as H. G. Khorana, who shared the Nobel Prize in 1968. After obtaining his PhD, Roger Kornberg, currently a professor at Stanford University, did research with Aaron Klug (Nobel Prize in Chemistry, 1982). None of these Nobel laureates did research on adrenaline. However, immediately after winning his Nobel Prize, a tired Roger Kornberg had this exchange with a journalist: "You'll be running on adrenaline for days to come, I imagine," the journalist remarked, and Kornberg responded, "I'm sure it's extraordinary news." On the other hand, Vladimir Prelog (Nobel Prize in Chemistry in 1975) believed that his own adrenaline level was unusually low.

22. A. Kornberg, "Remembering Our Teachers," *Journal of Biological Chemistry* 276 (2001): 3–11.

Glossary

Acetylcholine. The major neurotransmitter in the parasympathetic nervous system as well as in other locations in the nervous system.

Addison's disease. A disease that occurs as a consequence of loss of function of the adrenal glands. The major symptoms are primarily due to deficiency of the hormones cortisol and aldosterone, which are normally produced in the adrenal cortex.

Adenylyl cyclase (AC). The enzyme that synthesizes cAMP.

Adrenal cortex. See Adrenal glands.

Adrenal glands. Pair of glands located above each kidney. Adrenal glands are made up of an outer layer called the cortex, which surrounds the inner adrenal medulla. The principal hormones of the cortex are cortisol and aldosterone (steroids); adrenaline is the major hormone secreted by the medulla.

Adrenal medulla. See Adrenal glands.

Adrenalectomy. The surgical removal of an adrenal gland.

Adrenaline. The principal hormone of the adrenal medulla. *Adrenaline* is synonymous with *epinephrine*.

Adrenergic. An adjective indicating that a process in some way involves adrenaline and related drugs, the sympathetic nervous system, or receptors that respond to adrenaline.

Agonist. In pharmacology, a drug that has the capacity to activate a receptor; the receptor in turn stimulates a specific signal transduction pathway.

Angina. Chest discomfort due to inadequate blood flow to heart muscle. Angina typically occurs in patients with narrowed coronary arteries.

Antagonist. In pharmacology, a drug that binds to a receptor but does not have the capacity to activate the signaling pathway used by the receptor. On the other hand, by occupying the receptor, an antagonist prevents agonists from activating the same receptor.

Asthma. A common lung disease associated with reversible narrowing of small airways deep in the lungs. Narrowing of these airways makes it hard for asthmatics to move air in and out of the lungs. Adrenaline and related drugs help relax the smooth muscle in these airways, allowing them to become temporarily wider.

Autonomic nervous system. A system composed of the sympathetic and parasympathetic nervous systems. The brain uses these nerves to control involuntary functions of organs throughout the body; these systems often have opposing actions on target organs. For example, sympathetic nerve activity increases heart rate, while parasympathetic nerves activity slows beating.

cAMP. Adenosine 3',5'-cyclic monophosphate, a chemical synthesized from adenosine triphosphate (ATP) by the enzyme adenylyl cyclase. cAMP is a second messenger that activates the enzyme protein kinase A (PKA), which in turn phosphorylates target proteins.

Cardiac arrest. The sudden loss in capacity of the heart to pump blood, which causes unconsciousness due to lack of blood flow to the brain. Cardiac arrests can occur with a heart attack or with electric instability in the heart. A cardiac arrest is fatal unless cardiopulmonary resuscitation is started within minutes.

Cardiopulmonary resuscitation. Efforts aimed at restoring the function of the heart after a cardiac arrest.

Catecholamines. Substances that include adrenaline, noradrenaline, and dopamine.

Coronary arteries. Arteries that supply blood to the heart muscle. The commonest form of coronary artery disease is atherosclerosis, which can narrow the arteries or produce a heart attack.

Cortisol. A major hormone of the adrenal cortex; essential for life.

Endocrine glands. Specialized organs that secrete hormones. Major endocrine glands include the pituitary, the thyroid, the adrenals, the gonads, and the part of the pancreas that secretes insulin and glucagon. Endocrinology involves the study of endocrine glands.

Enzymes. Large protein molecules that catalyze (speed up) chemical reactions. Adenylyl cyclase is an enzyme that converts ATP to cAMP. Specialized enzymes called kinases have the capacity to phosphorylate (add a phosphate chemical group to) proteins.

Extract. Animal tissues that have been broken up in order to release the contents of their cells.

Fight-or-flight response. A term summarizing the integrated, autonomic reactions that may occur in response to threats. Many of the physiological changes in these circumstances are mediated by adrenaline and the sympathetic nervous system.

G proteins. A family of proteins important in signal transduction. When activated by a receptor on the cell surface, specific G proteins bind GTP and in turn activate biochemical processes within cells.

Glucagon. A hormone that activates specific glucagon receptors which in turn lead to increased concentrations of cAMP in cells.

Glycogen. A large molecule made up of many molecules of glucose chemically strung together. When energy requirements increase, glycogen stores in the liver and skeletal muscles are broken down to liberate glucose. Glycogen is similar to starches (carbohydrates) found in plants.

Heart attack. Usually occurs after a blood clot suddenly blocks the flow of blood in an atherosclerotic coronary artery. Deprivation of blood flow to heart muscle leads to massive death of heart cells, which can cause failure of the heart as a pump or lead to sudden death if the electrical system of the heart misfires.

Hormones. Chemicals typically synthesized in endocrine glands that are released into the blood in order to regulate the activity of distant organs. Adrenaline is the principal hormone of the adrenal medulla. Other examples of major hormones include thyroxine, made by the thyroid gland, and estrogens and testosterone, made primarily by the ovaries and testicles, respectively.

Ischemia. A pathological state caused by inadequate blood flow to tissues, leading to deficient supply of oxygen and nutrients such as glucose that are needed to support the metabolic needs of the tissue. Narrowing of the coronary arteries, typically by atherosclerosis, can cause ischemia of the heart muscle, which characteristically occurs when the work of the heart increases with exercise or emotional stress.

Neuron. The fundamental individual cell that makes up the nervous system.

Neurotransmission. The release of chemicals by nerve endings that then diffuse across the very narrow gap to cells in target organs, where they activate specific receptors. Noradrenaline and acetylcholine are the major neurotransmitters that activate target tissues of the sympathetic and parasympathetic nervous systems, respectively.

Nobel Prize. Alfred Nobel's estate established prizes in Physics, Chemistry, Physiology or Medicine, Literature, and Peace that were first awarded in 1901. These prizes are the most prestigious awards in their fields; winners are frequently referred to as Nobel laureates. The current financial value of a prize is just over $1 million. Nobel Prize in Medicine or Physiology is abbreviated to "Nobel Prize in Medicine" throughout the book.

Noradrenaline. The main transmitter in the sympathetic nervous system. Its structure is very similar to adrenaline, missing just one small chemical group in comparison.

Parasympathetic nervous system. A branch of the autonomic nervous system that innervates many major organs of the body, often slowing down their function. Neurons in this branch of the nervous system release acetylcholine near target cells which mediates their effects.

Partial agonist. In pharmacology, a drug that can partially activate responses mediated by a receptor. In other words, a partial agonist on its own typically induces a smaller effect than a full agonist such as adrenaline. Partial agonists prevent full agonist drugs from activating these receptors and consequently diminish their effects.

Pharmacology. A very broad and deep scientific discipline primarily concerned with the actions of drugs and analogous substances.

Pheochromocytoma. Tumors that arise from specialized nerve-like chromaffin cells primarily found in the adrenal medulla. These tumors can secrete enormous quantities of adrenaline and noradrenaline, which can cause headaches and palpitations of the heart along with high blood pressure.

Phosphorylation. The addition of a phosphate chemical group to another molecule. The phosphorylation of enzymes and other proteins can markedly change their activity or function.

Physiology. A biomedical science that investigates the function of living organisms. Experiments may involve intact life-forms or their components, such as organs or cells.

Postganglionic neuron. A nerve cell that carries signals from a ganglion to a target organ.

Preganglionic neuron. A nerve cell that carries signals from the central nervous system to a ganglion.

Protein kinase. An enzyme capable of phosphorylating (adding a phosphate group) to another protein.

Radial artery. An artery that supplies blood to the hands. The radial artery is the conventional artery used for checking a patient's pulse to measure heart rate.

Receptor. In this book the term *receptors* refers to large proteins on the cell surface that recognize and bind specific drugs or hormones. There are three main classes of adrenergic receptors: α_1, α_2, and β. Each of the main classes of adrenergic receptors is composed of three subtypes.

Rhodopsin. The protein that detects photons entering the eye.

Selective drug. A drug that preferentially interacts with one type of adrenergic receptor compared to another type of adrenergic receptor is considered selective. In other words, a selective drug has the capacity to discriminate between at least two receptors.

Signal transduction. The sequence of biochemical events that occurs after a drug or hormone activates its receptor.

Sympathetic nervous system (SNS). A branch of the autonomic nervous system that innervates major organs in the body. For example, neurons in the SNS release noradrenaline near cells in the heart, causing the heart to beat harder and faster. Adrenaline has many effects similar to those of the SNS.

Sympatholytic drug. A drug that interferes with some or all the effects of activating the sympathetic nervous system. Many sympatholytic drugs are antagonists that bind to adrenergic receptors. Other sympatholytic drugs do not bind to adrenergic receptors but inhibit the expected actions of adrenaline or noradrenaline by other mechanisms.

Sympathomimetic drug. A drug that mimics some or all the effects of activating the sympathetic nervous system. Many sympathomimetic drugs bind to adrenergic receptors and activate them. Other sympathomimetic drugs work indirectly, by potentiating the effects of endogenous adrenaline or noradrenaline.

Further Reading

GENERAL WORKS

Brunton, L., B. Chabner, and B. Knollman. 2011. *Goodman and Gilman's The Pharmacological Basis of Therapeutics*. 12th ed. New York: McGraw Hill Medical.

Gomperts, R. D., I. M. Kramer, and P. E. R. Tatham. 2009. *Signal Transduction*. 2nd ed. Amsterdam: Elsevier.

Perez, D. M. 2006. *The Adrenergic Receptors in the 21st Century*. Totowa, NJ: Humana Press.

I. THE GOLDILOCKS PRINCIPLE

Burn, J. Harold. 1971. *The Autonomic Nervous System: For Students of Physiology and of Pharmacology*. Oxford: Blackwell Scientific.

Bybee, K. A., and A. Prasad. 2008. "Stress-Related Cardiomyopathy Syndromes." *Circulation* 118:397–409.

Cronin, A. 2008. "Charles Sugrue, M.D., of Cork (1775–1816) and the First Description of a Classical Medical Condition: Phaeochromocytoma." *Irish Journal of Medical Science* 177:171–175.

Frazier, Ian. 2004. "Legacy of a Lonesome Death." *Mother Jones* 29, Nov./Dec.

Messerli, F. H., K. R. Loughlin, A. W. Messerli, and W. R. Welch. 2007. "The President and the Pheochromocytoma." *American Journal of Cardiology* 99:1325–1329.

Neumann, H., et al. 2007. "Evidence of MEN-2 in the Original Description of Classic Pheochromocytoma." *New England Journal of Medicine* 357:1311–1315.

Phelps, M. William. 2003. *Perfect Poison*. New York: Pinnacle Books.

Sharkey, S. W., et al. 2011. "Why Not Just Call It Tako-Tsubo Cardiomyopathy: A Discussion of Nomenclature." *Journal of the American College of Cardiology* 57:1496–1497.

Tormey, W. 2002. "Milestones in the Evolution of Phaeochromocytoma." *Journal of the Irish Colleges of Physicians and Surgeons* 31:222–231.

Van Heerden, J. A. 1982. "First Encounters with Pheochromocytoma. The Story of Mother Joachim." *American Journal of Surgery* 144:277–279.

Welbourn, R. B. 1987. "Early Surgical History of Phaeochromocytoma." *British Journal of Surgery* 74:594–596.

2. RULED BY GLANDS

Addison, Thomas. 1855. *On the Constitutional and Local Effects of Disease of the Supra-renal Capsules.* London: Samuel Highley.

Aminoff, Michael J. 2010. *Brown-Séquard: An Improbable Genius Who Transformed Medicine.* New York: Oxford University Press.

Bernard, Claude. 1957. *An Introduction to the Study of Experimental Medicine.* New York: Dover Publications. English translation of Bernard's book, originally published in 1865.

Bishop, P. M. F. 1950. "The History of the Discovery of Addison's Disease." *Proceedings of the Royal Society of Medicine* 43:35–42.

Carmichael, S. W. 1989. "The History of the Adrenal Medulla." *Reviews in the Neurosciences* 2:83–99.

Dirckx, J. H. 2001. "The Synthetic Genitive in Medical Eponyms: Is It Doomed to Extinction?" *Panace* 2:15–24.

Eknoyan, G. 2004. "Emergence of the Concept of Endocrine Function and Endocrinology." *Advances in Chronic Kidney Disease* 11:371–376.

Grande, Francisco, and Maurice B. Vissher, eds. 1967. *Claude Bernard and Experimental Medicine.* Cambridge, Mass.: Schenkman.

Lenard, A. 1951. "The History of Research on the Adrenals; 1563–1900." *Journal of the History of Medicine and Allied Sciences* 6:496–505.

Loughlin, K. R. 2002. "John F. Kennedy and His Adrenal Disease." *Urology* 59:165–169.

Løvås, K. and Husebye, E. S. 2005 "Addison's Disease." *Lancet* 365:2058–2061.

Medvei, Victor C. 1993. *The History of Clinical Endocrinology.* Pearl River, N.Y.: Parthenon.

Olmsted, J. M. D. 1946. *Charles-Édouard Brown-Séquard: A Nineteenth Century Neurologist and Endocrinologist.* Baltimore: Johns Hopkins University Press.

Rolleston, Humphry D. 1936 *The Endocrine Organs in Health and Disease: With an Historical Review.* London: Oxford University Press.

Sawin, C. T. 2001. "The Invention of Thyroid Therapy in the Late Nineteenth Century." *Endocrinologist* 2:1–3.

Stewart G. N. 1924. "Adrenalectomy and the Relation of the Adrenal Bodies to Metabolism." *Physiological Reviews* 4:163–190.

Wilson, L. G. 1984. Internal Secretions in Disease: The Historical Relations of Clinical Medicine and Scientific Physiology. *Journal of the History of Medicine and Allied Sciences* 39:263–302.

3. A COUNTRY DOCTOR'S REMARKABLE DISCOVERY

Barcroft, H., and J. F. Talbot. 1968. "Oliver and Schäfer's Discovery of the Cardio-vascular Action of Suprarenal Extract." *Postgraduate Medical Journal* 44:6–8.

Borell, M. 1978. "Setting the Standards for a New Science: Edward Schäfer and Endocrinology." *Medical History* 22:282–290.

Dale, H. H. 1948. Accident and Opportunism in Medical Research. *British Medical Journal* 2:451–455.

Davenport, H. W. 1991. "Early History of the Concept of Chemical Transmission of the Nerve Impulse." *Physiologist* 34:129, 178–190.

Goldstein, David. 2006. *Adrenaline and the Inner World: An Introduction to Scientific Integrative Medicine.* Baltimore: Johns Hopkins University Press.

Greek, C. Ray, and Jean Swingle Greek. 2000. *Sacred Cows and Golden Geese: The Human Cost of Experiments on Animals.* New York: Continuum.

Henderson, John. 2005. *A Life of Ernest Starling.* Oxford: Oxford University Press.

Henriksen, J. 2003. "Starling, His Contemporaries and the Nobel Prize." *Scandinavian Journal of Clinical and Laboratory Investigation* 238:1–59.

Henriksen, J. H., and O. B. Schaffalitzky de Muckadell. 2000. "Secretin, Its Discovery and the Introduction of the Hormone Concept." *Scandinavian Journal of Clinical and Laboratory Investigation* 60:463–472.

Langlois, Paul. 1897. "Les Capsules Surrénales." Paris: Ancienne Librairie Germer Baillière et Cie.

Oliver, George. 1895. *Pulse-Gauging: A Clinical Study of Radial Measurement and Pulse-Pressure.* London: H. K. Lewis.

Rolleston, Humphry D. 1936 *The Endocrine Organs in Health and Disease: With an Historical Review.* London: Oxford University Press.

Schäfer, E. A. 1908. "Present Condition of Our Knowledge Regarding the Function of the Suprarenal Capsules." *Lancet* i:1277–1281.

Wilson, J. D. 2005. "The Evolution of Endocrinology." *Clinical Endocrinology* 62:389–396.

Young, F. G. 1970. The Evolution of Ideas about Animal Hormones. In *The Chemistry of Life: Lectures on the History of Biochemistry*, ed. Joseph Needham. Cambridge: Cambridge University Press.

4. FINDING A NEEDLE IN A HAYSTACK

Abel, J. J. 1926. "Arthur Robertson Cushny and Pharmacology." *Science* 63:507–515.

Altman, Lawrence K. 1987. *Who Goes First? The Story of Self-Experimentation in Medicine.* New York: Random House.

Americans of Japanese Ancestry and the United States Constitution. 1987. San Francisco: National Japanese American Historical Society.

Aronson, Jeffrey. 2000. "Where Name and Image Meet—The Argument for Adrenaline." *British Medical Journal* 320:506–509.

Bennett, J. W. 2001. Adrenalin and Cherry Trees. *Modern Drug Discovery* 4:47–51.

———. 2002. "In Search of Dr. Jokichi Takamine and the Origins of Industrial Mycology." *Inoculum: Newsletter Supplement to Mycologia* 53(6):6–9.

———. 2009. "*Aspergillus*: A Primer for the Novice." *Medical Mycology* suppl. I:S5–S12.

Dale, H. H. 1933. "Academic and Industrial Research in the Field of Therapeutics." *Science* 77:521–527.

Davenport, H. W. 1982. Epinephrin(e). *Physiologist* 25:76–82.

De Mille, Agnes. 1978. *Where the Wings Grow.* Garden City, N.Y.: Doubleday.

Ishida, Mitsuo, ed. 2007. *Jokichi Takamine: The Man Who Gave "Adrenaline" to the World: English Translation of the Panels Exhibited at the National Science Museum, Tokyo, Japan, December 10, 2004–January 10, 2005 in Commemoration of the 150th Anniversary of Takamine's Birth.* Japanese Scientists and Technologists Series 1. Yokohama: Research Conference on Modern Creative Japanese Scientists.

Kawakami, K. K. 1928. *Jokichi Takamine: A Record of His American Achievements.* New York: William Edwin Rudge.

McClellan, A. 2005. *The Cherry Blossom Festival: Sakura Celebration.* Boston: Bunker Hill Publishing.

Parascandola, John. 1992. *The Development of American Pharmacology: John J. Abel and the Shaping of a Discipline.* Baltimore: Johns Hopkins University Press.

Sneader, W. 2001. "The Discovery and Synthesis of Epinephrine." *Drug News Perspectives* 14:539–543.

Swann, J. P. 1988. *Academic Scientists and the Pharmaceutical Industry: Cooperative Research in Twentieth-Century America.* Baltimore: Johns Hopkins University Press.

Takamine, J. 1902. "The Blood Pressure Raising Principle of the Suprarenal Gland." *Journal of the American Medical Association* 38:153–155.

Tansey, E. M. 1995. "What's in a Name? Henry Dale and Adrenaline, 1906." *Medical History* 39:459–476.

Voegtlin, C. 1939. "John Jacob Abel, 1857–1938." *Journal of Pharmacology and Experimental Therapeutics* 67:373–406.

Yamashima, T. 2003. "Jokichi Takamine (1854–1922), the Samurai Chemist, and His Work on Adrenalin." *Journal of Medical Biography* 11:95–102.

Yamashita, A. 2002. "Research Note on Adrenaline by Keizo Uenaka in 1900." *Biomedical Research* 23:1–10.

5. ADRENALINE ZIPS FROM BENCH TO BEDSIDE

Cockcroft, D. W. 1999. "Pharmacologic Therapy for Asthma: Overview and Historical Perspective." *Journal of Clinical Pharmacology* 39:216–222.

Crile, George. 1947. *An Autobiography, Volume 1.* Philadelphia: J. B. Lippincott.

Doll, Richard. 1998. "Controlled Trials: The 1948 Watershed." *British Medical Journal* 317:1217–1220.

Johnston, S. L., J. Unsworth, and M. M. Gompels. 2003. "Adrenaline Given outside the Context of Life Threatening Allergic Reactions." *British Medical Journal* 326:589–590.

Olasveengen, T. M. 2009. "Intravenous Drug Administration during Out-of-Hospital Cardiac Arrest." *Journal of the American Medical Association* 302:2222–2229.

Takamine, J., et al. 1901. *Adrenalin: Considered from a Historical, Chemical, Pharmaceutical and Therapeutic Standpoint.* Detroit: Parke, Davis.

Thevasagayam, M., et al. 2007. "Does Epinephrine Infiltration In Septoplasty Make Any Difference?" *European Archives of Oto-Rhino-Laryngology* 264:1175–1178.

6. MIND THE GAP

Axelrod, J. 1988. "An Unexpected Life in Research." *Annual Review of Pharmacology* 28:1–23.

Bennett, Max R. 2001. *History of the Synapse.* Amsterdam: Harwood.

Bisset, G. W., and T. V. P. Bliss. 1997. "Wilhelm Siegmund Feldberg, C. B. E." *Biographical Memoirs of Fellows of the Royal Society* 43:144–170.

Blaschko, H. K. F. 1980. "My Path to Pharmacology." *Annual Review of Pharmacology and Toxicology* 20:1–14.

Church, Roy, and E. M. Tansey. 2007. *Burroughs Wellcome and Co.* Lancaster: Crucible Books.

Dale, H. H. 1958. "Autobiographical sketch." *Perspectives in Biology and Medicine* 1:125–137.

———, 1961. "Thomas Renton Elliott." *Biographical Memoirs of Fellows of the Royal Society* 7:53–74.

Davenport, H. W. 1991. "Early History of the Concept of Chemical Transmission of the Nerve Impulse." *Physiologist* 34:129, 178–190.

Feldberg, W. S. 1982. *Fifty Years On: Looking Back on Some Developments in Neurohumoral Physiology.* Liverpool: Liverpool University Press.

Fick, G. R. 1987. "Henry Dale's Involvement in the Verification and Acceptance of the Theory of Neurochemical Transmission: A Lady in Hiding." *Journal of the History of Medicine and Allied Sciences* 42:467–485.

Fletcher, W. M. 1926. "John Newport Langley. In Memoriam." *Journal of Physiology* 61:1–16.

Hoffman, Brian B., and Palmer Taylor. 2001. "Drugs Acting at Synaptic and Neuroeffector Junctional Sites." In *Goodman and Gilman's The Pharmacologic Basis of Therapeutics*, 10th ed. Edited by J. Hardman and L. Limbird. New York: McGraw-Hill.

Kanigel, Robert. 1986. *Apprentice to Genius: The Making of a Scientific Dynasty.* New York: Macmillan.

Loewi, O. 1960. "An Autobiographical Sketch." *Perspectives in Biology and Medicine* 4:168–190.

Medawar, Jean, and David Pyke. 2000. *Hitler's Gift: The True Story of the Scientists Expelled by the Nazi Regime.* New York: Arcade.

Rapport, Richard. 2005. *Nerve Endings: The Discovery of the Synapse.* New York: W. W. Norton.

Rubin, R. P. 2007. "A Brief History of Great Discoveries in Pharmacology: In Celebration of the Centennial Anniversary of the Founding of the American Society of Pharmacology and Experimental Therapeutics." *Pharmacological Reviews* 59:289–359.

Snyder, S. H. 2006. "Turning off Neurotransmitters." *Cell* 125:12–15.

Valenstein, Elliot S. 2005. *The War of the Soups and the Sparks: The Discovery of Neurotransmitters and the Dispute over How Nerves Communicate.* New York: Columbia University Press.

Von Euler, U. S. 1971. "Pieces in the Puzzle." *Annual Review of Pharmacology* 11:1–13.

7. HOW ADRENALINE STIMULATES CELLS

Birnbaumer, L. 1999. "Martin Rodbell (1925–1998)." *Science* 283:1656.

———. 2007. "The Discovery of Signal Transduction by G Proteins. A Personal Account and an Overview of the Initial Findings and Contributions That Led to Our Present Understanding." *Biochimica et Biophysica Acta* 1768:756–771.

Bourne, Henry R. 2009. *Ambition and Delight: A Life in Experimental Biology.* N.p.: Xlibris.

Cohen, P. 2002. "The Origins of Protein Phosphorylation." *Nature Cell Biology* 4:E127–E130.

Cori, C. F. 1969. "The Call of Science." *Annual Review of Biochemistry* 38:1–21.

———. 1978. "Earl W. Sutherland." *National Academy of Sciences Biographical Memoirs* 2:319–350.

"Cross Talk: Interview with Al Gilman." 2001. *Molecular Interventions* 1:14–21.

Gilman, A. G. 2011. "Silver Spoons and Other Personal Reflections." *Annual Review of Pharmacology and Toxicology* 52:1–19.

Gravest, J. D., and E. G. Krebs. 1999. "Protein Phosphorylation and Signal Transduction." *Pharmacology and Therapeutics* 82:111–121.

Krebs, E. G. 1998. "An Accidental Biochemist." *Annual Review of Biochemistry* 67:xiii–xxxii.

Milligan, G., and E. Kostenis. 2006 "Heterotrimeric G-Proteins: A Short History." *British Journal of Pharmacology* 147:S46–S55.

Ochoa, S., and H. M. Kalckar. 1958. "Gerty Cori, Biochemist." *Science* 17:16–17.

Ritchie, M. 1996. "Alfred Gilman." *Biographical Memoirs. National Academy of Sciences* 70:59–81.

Robison, G. A., R. W. Butcher, and E. W. Sutherland. 1971. *Cyclic AMP.* New York: Academic Press.

Rodbell, M. 1991. "The Beginnings of an Endocrinologist." *Endocrinology* 129:2807–2808.

Sinding, C. 1996. "Literary Genres and the Construction of Knowledge in Biology: Semantic Shifts and Scientific Change." *Social Studies of Science* 26:43–70.

Woolf, P. K. 1975 "The Second Messenger: Informal Communication in Cyclic AMP Research." *Minerva* 13:349–373.

8. LOCK AND KEY

Ahlquist, R. P. 1948. "A Study of the Adrenotropic Receptors." *American Journal of Physiology* 153:586–600.

———. 1973. "Adrenergic Receptors: A Personal and Practical View." *Perspectives in Biology and Medicine* 17:119–122.

Barger, G., and H. H. Dale. 1910. "Chemical Action Structure and Sympathomimetic Action of Amines." *Journal of Physiology* 41:19–59.

Bennett, M. R. 2000. "The Concept of Transmitter Receptors: 100 Years On." *Neuropharmacology* 39:523–546.

Black, J. 2010. "A Life in New Drug Research." *British Journal of Pharmacology* 160 (suppl. 1):S15–S25.

Bosch, F., and L. Rosich. 2008. "The Contributions of Paul Ehrlich to Pharmacology: A Tribute on the Occasion of the Centenary of His Nobel Prize." *Pharmacology* 82:171–179.

Colquhoun, D. 2006. "The Quantitative Analysis of Drug-Receptor Interactions: A Short History." *Trends in Pharmacological Sciences* 27:149–157.

"Cross Talk: Interview with Sir James Black." 2001. *Molecular Interventions* 4:139–142.

Cuello, A. C. 1998. "Mark Nickerson (1916–1998)." *Trends in Pharmacological Sciences* 19:443–444.

Dale, H. H. 1940. "George Barger. 1878–1939." *Obituary Notices of Fellows of the Royal Society* 3:63–85.

Hill, S. J. 2006. "G-Protein-Coupled Receptors: Past, Present and Future." *British Journal of Pharmacology* 147:S27–S37.

Kresge, K., R. D. Simoni, and R. L. Hill. 2007. "Modeling β-Adrenergic Receptor Activation: The Work of Robert J. Lefkowitz." *Journal of Biological Chemistry* 282:e10–e11.

Lefkowitz, R. J. 2007. "Seven Transmembrane Receptors: Something Old, Something New." *Acta Physiologica* 190:9–19.

Parascandola, J. 1975. "Arthur Cushny, Optical Isomerism, and the Mechanism of Drug Action." *Journal of the History of Biology* 8:145–165.

Parascandola, J., and R. Jasensky. 1974. "Origins of the Receptor Theory of Drug Action." *Bulletin of the History of Medicine* 48:199–220.

Philipp, M., and L. Hein. 2004 "Adrenergic Receptor Knockout Mice: Distinct Functions of 9 Receptor Subtypes." *Pharmacology and Therapeutics* 101:65–74.

Prüll, C. R., A. H. Maehle, and R. F. Halliwell. 2009. *A Short History of the Drug Receptor Concept.* Basingstoke: Palgrave Macmillan.

Quirk, V. 2006. "Putting Theory into Practice: James Black, Receptor Theory, and the Development of the Beta-Blockers at ICI, 1958–1978." *Medical History* 50:62–92.

Rang, H. P. 2006. "The Receptor Concept: Pharmacology's Big Idea." *British Journal of Pharmacology* 147:S9–S16.

Shanks, R. G. 1984. "The Discovery of Beta-Adrenoceptor Blocking Drugs." *Trends in Pharmacological Sciences* 5:405–409.

Snyderman, R. 2011. "2011 Association of American Physicians George M. Kober Medal: Introduction of Robert J. Lefkowitz." *Journal of Clinical Investigation* 121:419–4200.

Vos, R. 1991. *Drugs Looking for Diseases*. Dordrecht: Kluwer.

William, R. 2010. "Robert Lefkowitz: Godfather of G Protein-Coupled Receptors." *Circulation Research* 106:812–814.

9. NEW DRUGS FROM OLD MOLECULES

Abraham, J., and C. Davis. 2006. "Testing Times: The Emergence of the Practolol Disaster and Its Challenge to British Drug Regulation in the Modern Period." *Social History of Medicine* 19:127–147.

Bristow, M. R. 2011. "Treatment of Chronic Heart Failure with β-Adrenergic Receptor Antagonists." *Circulation Research* 109:1176–1194.

Chen, K. K. 1981. "Two Pharmacological Traditions: Notes from Experience." *Annual Review of Pharmacology and Toxicology* 21:1–6.

Chen, K. K., and C. F. Schmidt. 1930. "Ephedrine and Related Substances." *Medicine* 9:1–117.

Davis, E., et al. 2008. "The Rush to Adrenaline: Drugs in Sport Acting on the Adrenergic System." *British Journal of Pharmacology* 154:584–597.

Fitch, K. D. 2006. "β_2-Agonists at the Olympic Games." *Clinical Reviews in Allergy and Immunology* 31:259–268.

Frishman, W. H. 2008. "β-Adrenergic Blockers: A 50-Year Historical Perspective." *American Journal of Therapeutics* 15:565–576.

Konzett, H. 1981. "On the Discovery of Isoprenaline." *Trends in Pharmacological Sciences* 2:47–49.

Kruse, P., et al. 1986. "Beta-Blockade Used in Precision Sports: Effect on Pistol Shooting Performance." *Journal of Applied Physiology* 61:417–420.

Pearce, Neil. 2007. *Adverse Reactions: The Fenoterol Story*. Auckland: Auckland University Press.

Persson, C. G. A. 1995. "Astute Observers Discover Anti-Asthma Drugs." *Pharmacology and Toxicology* 77 (suppl. 3):7–15.

Rasmussen, Nicholas. 2008 *On Speed: The Many Lives of Amphetamine*. New York: New York University Press.

Rau, J. L. 2000. "Inhaled Adrenergic Bronchodilators: Historical Development and Clinical Application." *Respiratory Care* 45854–45863.

Sears, M. R., and J. Lotvall. 2005. "Past, Present and Future—β_2-Adrenoceptor Agonists in Asthma Management." *Respiratory Medicine* 99:152–170.

Sneader, W. 2001. "Epinephrine Analogues." *Drug News Perspectives* 14:491–494.

———. 2005. *Drug Discovery: A History*. Chichester: John Wiley and Sons.

Soni, M. G., et al. 2004. "Safety of Ephedra: Lessons Learned." *Toxicology Letters* 150:97–110.

Stahle, H. 2000. "A Historical Perspective: Development of Clonidine." *Bailliere's Clinical Anaesthesiology* 14:237–246.

Tattersfield, A. E. 2006. "Current Issues with β_2-Adrenoceptor Agonists: Historical Background." *Clinical Reviews in Allergy and Immunology* 31:107–117.

10. ADRENALINE JUNKIES

Alberti, F. B. 2009. "Bodies, Hearts, and Minds: Why Emotions Matter to Historians of Science and Medicine." *Isis* 100:798–810.

Boissy, A. 1995. "Fear and Fearfulness in Animals." *Quarterly Review of Biology* 70:165–191.

Cannon, W. B. 1929. *Bodily Changes in Pain, Hunger, Fear and Rage: An Account of Recent Researches into the Function of Emotional Excitement*. 2nd ed. New York: D. Appleton.

———. 1942. "'Voodoo' Death." *American Anthropologist* n.s. 44:169–181.

Dale, H. H. 1947. "Walter Bradford Cannon (1871–1945)." *Obituary Notices of Fellows of the Royal Society* 5:407–423.

Damasio, Antonio. 1999. *Feeling of What Happens: Body and Emotion in the Making of Consciousness*. Orlando, Fla.: Harcourt Brace.

Davenport, H. W. 1981. "Signs of Anxiety, Rage, or Distress." *Physiologist* 24:1–5.

Engel, G. L. 1971. "Sudden and Rapid Death during Psychological Stress: Folklore or Folk Wisdom?" *Annals of Internal Medicine* 74:771–782.

———. 1977. "The Need for a New Medical Model: A Challenge for Biomedicine." *Science* 196:129–136.

Everson-Rose, S. A., and T. T. Lewis. 2005. "Psychosocial Factors and Cardiovascular Diseases." *Annual Review of Public Health* 26:469–500.

Flannery, F. T., et al. 2009. "Homicide by Fright: The Intersection of Cardiology and Criminal Law." *American Journal of Cardiology* 105:136–138.

James, G. D., and D. E. Brown. 1997. "The Biological Stress Response and Lifestyle: Catecholamines and Blood Pressure." *Annual Review of Anthropology* 26:313–335.

Kligfield, P. 1980. "John Hunter, Angina Pectoris and Medical Education." *American Journal of Cardiology* 45:367–369.

Kloner, R. A. 2006. "Natural and Unnatural Triggers of Myocardial Infarction." *Progress in Cardiovascular Diseases* 48:285–300.

Kreibig, S. D. 2010. "Autonomic Nervous System Activity in Emotion: A Review." *Biological Psychology* 84:394–421.

LeDoux, Joseph. 1996. *The Emotional Brain*. New York: Simon and Schuster.

Leeka, J., et al. 2010. "Sporting Events Affect Spectators' Cardiovascular Mortality: It Is Not Just a Game." *American Journal of Medicine* 123: 972–977.

Leor, J., et al. 1996. "Sudden Cardiac Death Triggered by an Earthquake." *New England Journal of Medicine* 334:413–419.

Ma, W., et al. 2011. "Stock Volatility as a Risk Factor for Coronary Heart Disease Death." *European Heart Journal* 32:1006–1011.

Samuels, M. A. 2007. "The Brain-Heart Connection." *Circulation* 116:77–84.

Selye, Hans. 1956. *The Stress of Life*. New York: McGraw-Hill.

Strike, P. C., and A. Steptoe. 2005. "Behavioral and Emotional Triggers of Acute Coronary Syndromes: A Systematic Review and Critique." *Psychosomatic Medicine* 67:179–186.

Wittstein, I. S., et al. 2005. "Neurohumoral Features of Myocardial Stunning Due to Sudden Emotional Stress." *New England Journal of Medicine* 352:539–548.

APPENDIX

Carlsson, A. 1987. "Perspectives on the Discovery of Central Monoaminergic Neurotransmission." *Annual Review of Neuroscience* 10:19–40.

Davenport, H. W. 1981. "George H. Whipple or How to Be a Great Man without Knowing Differential Equations." *Physiologist* 24:1–5.

Forssmann-Falck, R. 1997. "Werner Forssmann: A Pioneer of Cardiology." *American Journal of Cardiology* 79:651–660.

Friedman, S. G. 1988 "Alexis Carrel: Jules Verne of Cardiovascular Surgery." *American Journal of Surgery* 155:420–424.

Hargittai, Istvan. 2002. *Candid Science II: Conversations with Famous Biomedical Scientists.* London: Imperial College Press.

———. 2002. *The Road to Stockholm: Nobel Prizes, Science, and Scientists.* Oxford: Oxford University Press.

Hawgood, B. J. 2004. "Professor Bernardo Alberto Houssay, MD (1887–1971): Argentine Physiologist and Nobel Laureate." *Journal of Medical Biography* 12:71–76.

Heymans, C. 1967. "Pharmacology in Old and Modern Medicine." *Annual Review of Pharmacology* 7:1–14.

Huber, A. 1982. "W. R. Hess, the Ophthalmologist." *Experientia* 38:1397–1400.

Kendall, Edward C. 1971. *Cortisone: Memoirs of a Hormone Hunter.* New York: Charles Scribner's Sons.

Kornberg, A. 2001. "Remembering Our Teachers." *Journal of Biological Chemistry* 276:3–11.

Levi-Montalcini, Rita. 1988. *In Praise of Imperfection: My Life and Work.* New York: Basic Books.

Liddell, E. G. T. 1952. "Charles Scott Sherrington." *Obituary Notices of Fellows of the Royal Society* 8:241–270.

Morris, J. B., and W. J. Schirmer. 1990. "The 'Right Stuff': Five Nobel Prize–Winning Surgeons." *British Journal of Surgery* 77:944–953.

Moss, Ralph W. 1988. *Free Radical: Albert Szent-Györgyi and the Battle over Vitamin C.* New York: Paragon House.

Oliverio, A. 1994. "Daniel Bovet." *Biographical Memoirs of Fellows of the Royal Society* 39:60–70.

Reggiani, A. H. 2002 "Alexis Carrel, the Unknown: Eugenics and Population Research under Vichy." *French Historical Studies* 25:331–356.

Tierney, A. J. 2000. "Egas Moniz and the Origins of Psychosurgery: A Review Commemorating the 50th Anniversary of Moniz's Nobel Prize." *Journal of the History of the Neurosciences* 9:22–36.

Acknowledgments

I obtained importance assistance from librarians at Countway Library at Harvard Medical School, Lane Library at Stanford University, the Harrogate Library in England, and the Peoria Public Library in Illinois, and especially from Ms. Irmeli Kilburn in the Medical Library at the VA Boston Healthcare System. Access to some of the letters of Agnes de Mille in the New York Public Library for the Performing Arts was obtained with the permission from her son, Jonathan Prude. Richard Gelbke of the National Archives in New York City helped me obtain legal documents from the adrenaline patent lawsuit. Jennifer Blue of the U.S. Geological Survey provided information about the Cori Lunar Crater. Anne Pearson helped me obtain information from the Library of the Diocese of Tucson about Jokichi Takamine. Bonnie Coles of the Library of Congress helped me obtain several letters involving Helen Taft. Joan Bennett provided encouragement at the beginning of this project. I appreciate the support of Michael Fisher at Harvard University Press. An anonymous reviewer made many helpful recommendations.

Friends and colleagues provided encouragement with frank criticisms of individual chapters. My wife, Maggie, read every page of the manuscript through several drafts and patiently made helpful recommendations. I would like to thank senior leadership at the VA Boston Healthcare System for providing an intellectual environment that greatly facilitated my writing this book.

Credits

The epigraph in Chapter 2 is excerpted from *Collected Essays* by Aldous Huxley, copyright © 1923, 1925, 1926, 1928, 1929, 1930, 1931, 1934, 1937, 1941, 1946, 1949, 1950, 1951, 1952, 1953, 1954, 1955, 1956, 1957, 1958, 1959 by Aldous Huxley, and reprinted by permission of Georges Borchardt, Inc., for the Estate of Aldous Huxley.

In Chapter 4 a verse from the song "Silent E" by Tom Lehrer is quoted with permission from the author.

Quotation in Chapter 10 from *Brave New World* by Aldous Huxley, published by Chatto & Windus, is reprinted by permission of The Random House Group Limited; in Canada, copyright © 1932, 1960 Aldous Huxley, reprinted by permission of Random House Canada; in the United States, copyright 1932, renewed © 1960 by Aldous Huxley and reprinted by permission of HarperCollins Publishers; ebook rights in the United States and Canada are copyright © 1932, 1946 by Aldous Huxley, reprinted by permission of Georges Borchardt, Inc., on behalf of the Aldous and Laura Huxley Trust. All rights reserved.

The epigraph in Chapter 10 is excerpted from *Adrenalin Mother*, from *The Pill versus the Springhill Mine Disaster* by Richard Brautigan. Copyright © 1965 by Richard Brautigan, and reprinted by permission of Houghton Mifflin Harcourt Publishing Company. All rights reserved.

Quotation in Chapter 10 from the World War Hulk Gamma Files, 2007, Marvel Comics, is trademarked and © Marvel & Subs and reprinted by permission.

Quotations from Nobel lectures throughout the book are used with the kind permission of the Nobel Foundation.

Index

Abel, John Jacob: career of, 43, 49–51, 215n4, 216n13; discovery of adrenaline, 216n10; dispute with von Fürth, 45, 56; epinephrin, 44–45, 46, 59, 61, 63; insulin studies, 204, 248n13; recognition of, 58, 61, 206

Abel number, 216n13

Acceleransstoff, 89, 92–94. *See also* Noradrenaline

Acetaminophen (Tylenol), 102

Acetylcholine: assay, 91, 230n19; arterial dilation with, 197; atropine and, 241n3; blocking of, 129; muscle contraction and, 168, 251n3; in parasympathetic nervous system, 91, 97; properties of, 230n21; sweat glands and, 232n26; synthesis of, 88, 228n10; *Vagusstoff*, 89, 90–92

Acetylcholinesterase, 90, 229n15

Addison, Thomas, 22–24, 25, 29, 60, 189, 209n3

Addison's disease, 24–26, 33, 75–76, 209n3, 225n18. See Kennedy, John F

Adenosine 3', 5'-cyclic monophosphate. *See* Cyclic AMP

Adenosine-5'-triphosphate (ATP), 97, 110, 111, 116, 168, 232n26

Adenylyl cyclase (AC): activation of, 115, 137, 238n9; cAMP synthesis, 110, 111, 116–118, 120, 137–138; cyc⁻ cells, 118, 120–121; GTP and, 117, 122

Adrenal cortex, 21, 22, 36, 42, 76, 184, 255n3

Adrenalectomies, 25, 26

Adrenal glands: active substances in, 33, 34, 36, 54–55, 85; discovery of, 19, 20–21, 22, 27, 209n2; function of, 6, 19, 21–24, 26; sympathetic nervous system and, 14. *See also* Adrenal cortex; Adrenalin(e); Adrenal medulla

Adrenalin(e): absorption of, 37, 213n3, 225n15; adverse effects of, 5–6, 79–83; biosynthesis of, 98–100; chemical structure of, 46–48; discovery of, 1–3, 19, 42, 43, 58, 206, 216n10; early uses of, 66–71; lawsuit over, 61, 62–65; literature mentions of, 205; modern uses of, 77–78, 187; naming of, 44, 46, 56, 59, 60–61, 161, 177–178, 215n2, 253n11; purification of, 44–47, 57–58, 218n23; receptors for, 124–139; responses to, 160–185; rushes from, 165–169, 181; shock and awe of, 71–74; transformative effects of, 160–165. *See also* Excess adrenaline; Glucose; Noradrenaline; Pheochromocytomas

Adrenaline Addicts Anonymous (AAA), 179

Adrenal medulla: active substances in, 36, 41, 42, 43, 44, 61, 94, 189; initial messenger of, 111; noradrenaline in, 97, 100; structure of, 21, 22; synthesis in, 98. *See also* Chromaffin cells

Adrenergic receptors: biology of, 149–150; as framework for adrenaline analogs, 140, 151; polymorphisms in, 138–139; radioactively labeled, 134–135; structure, 136, 244n18; subtypes, 125, 139, 140, 150n22, 244n17. *See also* Alpha (α) adrenergic receptors; Beta (β) adrenergic receptors

Franklin Pierce University

00199688